15-MINUTE SCIENTIFIC THEORIES

ARCTURUS
This edition published in 2017 by Arcturus Publishing Limited
26/27 Bickels Yard, 151–153 Bermondsey Street,
London SE1 3HA

ISBN: 978-1-78428-606-4
AD005422UK

Printed in China

CONTENTS

Introduction
Why does science need theories?

Imagine you lived nearly 2,000 years ago in the town of Pompeii in Italy. One day, the nearby Mount Vesuvius blew tons of molten rock into the air and scorching winds devastated your town. If you survived, you might blame angry gods for this inexplicable event. Or you might wonder how the ground could pour out hot, semi-liquid rock and fiery air. The first option is the easiest.

If you assumed it was a natural rather than a supernatural event and looked for a physical cause, you would be taking a scientific approach. If you could suggest a plausible explanation that could fit all the facts, which did not

'Many besought the aid of the gods, but still more imagined there were no gods left, and that the universe was plunged into eternal darkness for evermore.'
Pliny the Younger, on the eruption of Mount Vesuvius, AD79

fail when tested against similar events, then you would have a scientific theory.

The first attempts at scientific thinking emerged in Ancient Greece, 2,500 years ago. Since then, science has moved in and out of favour. People have sometimes returned to superstitious or religious thinking that requires no analysis or investigation of causes and effects. At other times, scientific enquiry has been at the forefront of civic and cultural progress. The result is that science has progressed through leaps and bounds – or, rather, leaps and standstills.

Thoughts and theories

Science moves towards truth through a series of steps. Someone will have an idea, perhaps from observing some phenomenon. The idea will try to explain the phenomenon in a way that could formulate predictions or a general rule about other instances of the phenomenon. This attempt at explanation is a hypothesis.

For example, you might observe a bird such as a thrush bashing a snail against a stone. If you wanted to explain this, you might form the hypothesis that thrushes like the noise of snail shells smashing against stones and this is why they do it. You could then plan more observations; perhaps trap a thrush and some snails and see what happens. You would soon notice that the thrush eats the snail. This doesn't disprove your hypothesis – eating the snail might just be a

side benefit, or show a tendency to tidy up or hide the evidence. But it would open another avenue of exploration.

You might (rather unkindly, but this is science) deafen a thrush and see whether it persists in smashing snails when it can't hear the noise. I suspect that it would. The motivation for your particular thrush is not (now) to hear the noise of smashing shells; the activity might be instinctive. You could try feeding a thrush a pre-smashed snail. If it didn't eat it, this would support the idea that it's getting rid of the evidence of its smashing spree.

It probably wouldn't take many experiments for you to reject your original hypothesis and formulate a new one – that the thrush smashes a snail against a stone to break the shell so it can eat the snail meat.

This rather haphazard approach to investigating thrush behaviour reflects how science progressed in the early years of scientific enquiry. The procedure is much more rigorous and structured now, with all aspects of an experiment carefully controlled and separated, but the principle is still to think of a possible explanation, then test it and refine the explanation in the light of new evidence.

When a hypothesis is borne out by experiment or observation and can be used to make accurate predictions, it becomes a theory. A theory need not be accepted by everyone, but it must give a good, coherent explanation that has not been refuted by other evidence. The theory of evolution (see page 52) remains valid even though some people prefer a supernatural explanation for the existence of different species. The supernatural explanation is not supported by objective evidence, so it has no impact on the validity of the scientific theory.

A STROKE OF GENIUS, OR SLOW GENESIS

Some theories are the work of one person and their brilliant insight, challenging at a stroke the ideas that prevailed previously. Einstein's theories of relativity (see page 216) fall into this category. Other theories build up slowly, or rest on decades or centuries of previous research or thought carried out by many people. Gene theory (see page 232) is an example of a theory that relied on many pieces having been put in place before it became possible. There is usually one exceptionally creative mind behind the final synthesis, though. The history of science and its theories is a history of brilliant individuals drawing together the threads laid by their predecessors.

Theories change

Science differs from a belief system in that in science nothing is fixed forever. Every single current theory in modern science could be overturned and science would remain the same: a structure for understanding the universe.

A belief system - such as a religion - has fundamental statements that are taken as true. They don't need external proof - the conviction with which they are accepted is their sole justification. Indeed, asking for proof is often banned or at least frowned upon, as it reveals a lack of faith. In science, the opposite is the case. Something is only held as true if it can be supported by evidence, reliably reproduced or observed repeatedly; an individual's conviction is of no consequence. The only

THEORIES AND LAWS

Science has theories and it has laws. They are not the same thing, but sometimes they deal with the same phenomena. A theory tries to explain how and why things happen, to find the causes of phenomena. A law states what happens and allows calculations and predictions, but makes no attempt to explain why or how something happens. So the law of gravitation enables us to calculate the force acting between two bodies and how they will move, but it doesn't tell us *why* there is a force between them. Newton's theory of gravity explains how and why two bodies move in relation to each other.

meta-belief is that a rational explanation exists - even if it cannot be found - for everything that occurs in the natural universe. (This doesn't rule out the existence of a god: many scientists hold religious beliefs.)

Advancing through theories

Science advances through proposing and testing theories. Some are found to be right, some turn out to be not quite right and are adjusted, and some are found to be wrong (but that is also progress). In the following pages, we shall look at theories that are now accepted in scientific circles to be accurate representations of fact, some that have been overturned, and some that are still being discussed.

Understanding the theory of gravity has enabled space flight.

Chapter 1
Big Bang Theory

Has the universe always been there, throughout an infinity stretching backwards as well as forwards? Or was it created, either by a divinity or through some natural physical process?

Cosmic eggs

The question of where everything came from has troubled people for millennia. At first, the only approaches to it were conjectural or religious. People came up with myths to explain it or tried to find an explanation through philosophy; elements of both of these approaches are found in modern cosmological theory. A collection of Hindu hymns written in Sanskrit in India around 1500–1200BC, called the *Rig Veda*, describes a 'cosmic egg'. This was said to hold the entire content of the universe in a single, infinitely small point called a *Bindu*. The universe, and everything in it, expanded from this 'egg'.

THE IDEA

The universe began from a single, infinitely dense point that expanded massively to produce everything that exists, including space-time (the composite of space and time).

When the Belgian priest and astronomer Georges Lemaître first proposed what became known as the Big Bang theory in 1927 he also used the analogy of a 'cosmic egg'. But he came to his theory by a rather different route.

World(s) without end

The earliest record we have of rational ideas about how the universe might have started dates from the 6th century BC. The Greek philosopher Anaximander claimed that the 'boundless' has always existed and will always exist: 'all the heavens and the worlds within them' come from 'some boundless nature'. A century later, Anaxagoras proposed that the universe was a great mixture of ingredients divided into infinitely small fragments. The mix was set into a whirling motion which separated out the ingredients so that they clumped into the different types of matter and objects we see around us. His contemporary Empedocles suggested that two conflicting forces act to bring matter together and then tear it apart in a great, eternal cosmic cycle of creation and destruction. Later, the Stoics adopted a similar model of a repeating cycle. Again in the early 5th century BC, Parmenides argued that nothing can come into existence from nothing, nor disappear into nothing: the universe is

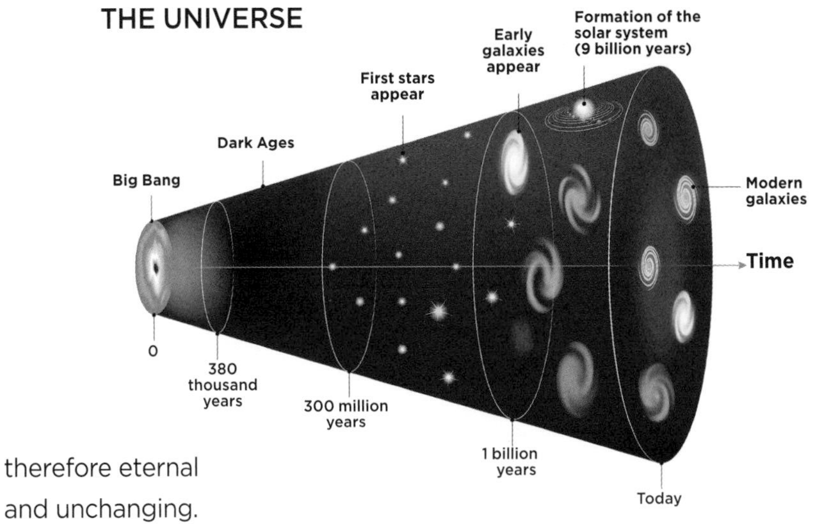

therefore eternal and unchanging.

Of these various models, that of an eternal universe became dominant in the West. It appealed to the Christian Church, to Judaism and to Islam as it allowed God to have initiated an eternal and perfect, unchanging creation in keeping with the teachings of the holy texts. For a long time, the universe was deemed to have no beginning or end.

Stasis challenged

The unchangeableness of the universe went largely unquestioned until a supernova (exploding star) appeared in 1572. The English astronomer Thomas Digges tried unsuccessfully to measure the distance to the 'new star'. The method he used was known to work when measuring the distance to objects that are relatively close; its failure showed that the star was beyond the orbit of the Moon – in the zone where things were not supposed to change. This spelled trouble; but not enough for people to reinvestigate the possible origins of the universe, outside a religious narrative, until the early 20th century. By then, it was 2,000 years since many of the arguments had been properly aired.

Einstein sets the ball rolling

In 1916, Austrian physicist Albert Einstein published his theory of general relativity (see page 216), using cutting-edge mathematics to propose a new model of how space, time and gravity work. It did not in itself address the origin of the universe, but the equations Einstein produced could be used to explore this.

Astronomer Tycho Brahe looking up at a supernova which appeared in 1572.

Personally, Einstein favoured a static universe. He believed the universe to be infinite but stable, neither growing nor shrinking. This was the prevailing view at the time. Einstein even added a 'cosmological constant' to his equations to force the universe to remain stable. Otherwise, the relativity equations showed that it changed size, which he considered to be quite clearly nonsense. Scientists working at the frontiers of knowledge are sometimes reluctant to let go of cherished beliefs and world views, but the need to fudge the equations is often a sign that something is fundamentally amiss.

All from nothing

And Einstein was, it seems, wrong. Many physicists and mathematicians played around in various ways with the equations Einstein laid out, some of them finding results that contradicted Einstein's own interpretations. One of them was Georges Lemaître.

Lemaître's work with the relativity equations suggested to him that the universe is constantly expanding. He also proposed that the most distant galaxies would be travelling away from the solar system the most quickly. Lemaître published his findings in 1927 in French and they went largely unnoticed because all the most influential astronomers only read journals published in

English or German. (The Russian astronomer Alexander Friedmann proposed the same idea in 1922, but did not pursue it.)

The Hooker Telescope at the Mount Wilson Observatory was used by Hubble.

Proof in the stars

At the same time as Lemaître was working in Europe, the American astronomer Edwin Hubble was examining distant galaxies at the Mount Wilson Observatory in California, USA. His findings in 1929 confirmed Lemaître's prediction: not only are galaxies moving away from Earth, but the most remote galaxies are moving away most quickly.

Two years later, in 1931, the English astronomer Arthur Eddington came across Lemaître's paper and arranged to have it translated into English. The same year, Lemaître announced at a meeting in London the fairly obvious conclusion that if everything is moving apart it had originally all been together. He proposed that the universe was once very small - infinitesimally small, in fact - and that the entire universe had expanded from a single point, a 'primaeval atom' or 'cosmic egg'.

Named in scorn

> *'The Big Bang, which is today posited as the origin of the world, does not contradict the divine act of creation; rather, it requires it.'*
> Pope Francis, 2014

It took a few years for Lemaître's theory to gain traction; it was unpopular to start with and even ridiculed in some quarters. The name by which it is now known, the Big Bang theory, comes from a sarcastic remark made by the English astronomer Fred Hoyle in 1949. But eventually Einstein and others came to accept it, and it is now the most widely accepted scientific paradigm for the origin of the universe, approved even by the Catholic Church.

Interfering pigeons?

The Big Bang theory had more to it than the similar proposals of the Ancient Greeks and Indian; observations of Hubble and Eddington supported it, as did Einstein's equations. But it remained a mathematical model until 1964, when two American radio astronomers, Arno Penzias and Robert Wilson, stumbled across the first physical evidence of the Big Bang. It might seem surprising that there can be evidence of an event which took place more than 13 billion years ago, before there were even any atoms to remain as witnesses. But information can be stored as energy (see page 246) and this is what Penzias and Wilson found.

The pair were working at Bell Laboratories in New Jersey, USA, on a super-sensitive new radio antenna. They were trying to remove all sources of interference and background noise, but even when they had eliminated all extraneous signals something remained. Their antenna still picked up faint background noise, 100 times the intensity they expected. Blaming pigeons nesting near the antenna, they evicted the birds, but the noise persisted. Its source seemed to be evenly spread across the sky, detectable both day and night.

Then Penzias and Wilson heard about work suggesting that radiation from the Big Bang might be detectable in space. They realized they had discovered just that - Cosmic Microwave Background Radiation (CMBR), the echo of the Big Bang, still travelling through space after 13.8 billion years.

Robert Wilson (left) and Arno Penzias by the radio antenna at Bell Laboratories, New Jersey.

Evenly spread

The Big Bang theory rests on Einstein's theory of general

relativity and an assumption called the Cosmological Principle. This states that if we look at the universe on a large enough scale, matter is evenly distributed and homogenous. Modern research supports the Cosmological Principle.

The shape of the universe

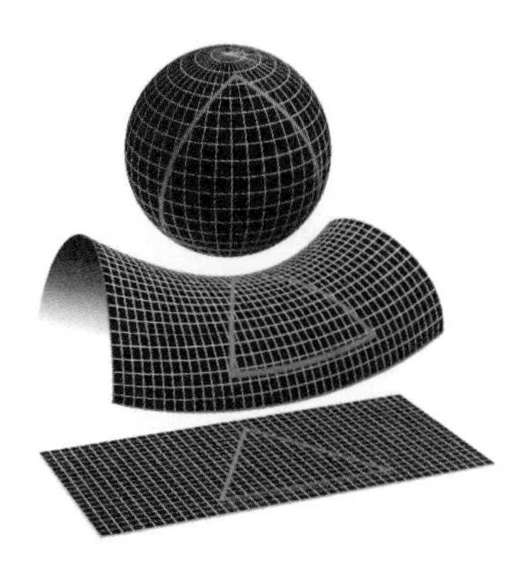

Assuming that matter is much the same everywhere, and Einstein's account of the interaction of gravity and space-time in the general theory of relativity is correct, there are only three possible shapes for the universe. It can be positively curved, like a ball; it can be negatively curved, like a saddle; or it can be flat.

The first of these gives a finite universe – starting from any point, if you travel far enough you will get back to where you started from. The other two models are potentially infinite. Cosmologists are not yet sure of the shape of the universe; it appears to depend on the average density of matter in the universe. If the average is more than the so-called critical density, the universe is positively curved. If it is less, the universe is negatively curved. If it is exactly the same, the universe is flat. The critical density is thought to be about six hydrogen atoms per cubic metre of space, and although we don't yet know the precise value of the average density of matter in space it seems to be close to this.

Theories about the shape and nature of the universe can only reasonably be applied to the visible universe. The nature of space

AS FAR AS WE CAN SEE . . .

Astronomers distinguish between the visible universe and the entire universe, which is potentially much larger, or even infinite. The visible universe extends to the same distance in all directions from Earth. It is limited by the speed of light – we can only 'see' (with telescopes, including those that detect radio or X-rays) as far as electromagnetic radiation can have travelled in the time that has passed since the start of the universe.

All electromagnetic radiation travels at the speed of light. It might seem, then, that if the universe is about 13.8 billion years old, we can only see 13.8 billion years in each direction. In fact, we can see further because the universe is expanding. It doesn't expand from the centre, but the space between stars and galaxies expands. This means a star that was once five million light years away could now be ten million light years away. Scientists estimate that the oldest electromagnetic radiation now reaching us has travelled 13.8 billion light years, but comes from objects that are now 46.5 billion light years away. The diameter of the visible universe is thought to be 93 billion light years.

might be very different beyond the horizon – we have no way of ever knowing. Accounts of the Big Bang and what came after it are based on observations of the visible universe, and could be completely wrong if the part we can see is not representative of the whole.

The Big Bang: not a bang, and possibly not big

The Big Bang was not an explosion. It was more that space, time and matter sprang into existence. If the universe is closed and finite, it probably did originate from an infinitely small point, known as a singularity. If it is infinite, its origins might also have been infinite – the universe simply blinked into existence everywhere – but then everywhere got progressively bigger.

The expansion of the universe is similarly difficult to envisage. It doesn't grow from the middle or the edge, and it's not expanding from any particular point. Instead, it's believed, new space-time appears all over the place. A helpful analogy is a ball of dough studded with raisins. As the dough is left to rise, it expands, and the raisins move further away from one another. Raisins that were close together in the dough are not separated as much by its expansion as raisins that were further apart to start with – but all move apart.

Where will it all end?

In the Hindu origins story, the universe eventually collapses once

A Hubble Space telescope image of our expanding universe.

more into the 'cosmic egg' before expanding again in an infinite cycle of expansions and contractions. Such a cycle remains a possible model under Big Bang theory.

Expansion could continue until (and beyond) the point when there is only an infinitely large, dilute soup of subatomic particles that have all been torn apart and scattered. Or expansion could go into reverse and the universe could collapse back into a point, perhaps beginning again with another Big Bang.

Which scenario is likely to be true depends on the density of matter. As matter produces a gravitational effect attracting other matter to it, gravity can slow down the expansion of the universe. However, recent evidence suggests that the rate of expansion is

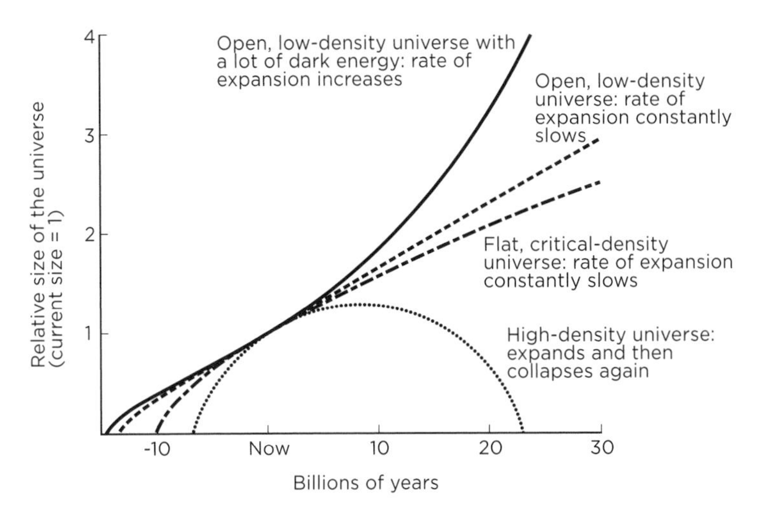

increasing rather than decreasing. That would put the universe on the top curve (see diagram above), heading rapidly for the dilute-soup fate.

DON'T GO THERE!

Big Bang theory deals with all of time and space, which is quite a large enough subject. As space-time started with the Big Bang, questions such as what was there before, or where did the stuff that is the universe come from before the Big Bang, or what is outside the universe, are meaningless. The Big Bang theory doesn't and can't address them.

Chapter 2

Theory of Spontaneous Generation

It's hard to see where some organisms come from – so in the past people assumed that they just appeared from nowhere.

THE IDEA

Some types of organism spring from inanimate matter under the right conditions.

Maggots from meat

Imagine you live hundreds of years ago, before the invention of the microscope. One day, you don't eat your lunch and forget to throw it away or feed it to a pig. A week later you find it's crawling with maggots. What would you assume from this? Possibly that old lunch generates maggots – that as the lunch decays, it 'makes' maggots? The maggots must have sprung from the inanimate matter that was your lunch. This is the theory of spontaneous generation.

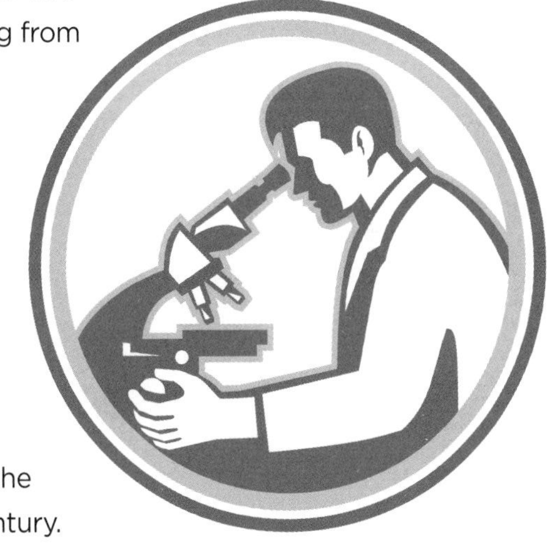

Although you may think the theory would be fairly easy to refute, it remained current for thousands of years and was only fairly conclusively dismissed in the second half of the 19th century.

No visible parents

The problem with maggots is that you never see any parent maggots. The realization that flies are the parents of maggots takes quite a leap, or at least some careful and dedicated observation. As people spent their time trying to kill flies or keep them away rather than studying them, it's not altogether surprising that they didn't make the connection early on.

> ***'Your Serpent of Egypt is bred now of your mud by the operation of your Sun: so is your Crocodile.'***
> William Shakespeare, *Antony and Cleopatra*, Act II, scene 7

Other animals which have no visible parents were explained in a variety of strange ways. All eels spawn far from Europe, in the Sargasso Sea (a region in the Atlantic off the coast of North America). For this reason, they are never seen to mate or spawn, and hatchlings are never seen in Europe. In the 4th century BC, the Greek

philosopher Aristotle concluded that eels came from earthworms. Later, in the 1st century AD, Pliny the Elder claimed they reproduced by budding, with the adults rubbing themselves against rocks to scrape off tiny new eels. And in the following century, Athenaeus believed that eels oozed out a kind of slime which settled on the mud of the river bank and generated new eels. (He also claimed anchovies were born out of sea-foam.)

In other cases, it is fairly easy to see how animals reproduce and fairly hard to see how anyone could miss the signs. Yet as late as the 17th century, when rigorous scientific research had been well under way for a while, the Flemish scientist Jan Baptiste van Helmont still believed that the 'fumes' from the bottom of a swamp gave rise

GROW-YOUR-OWN SCORPIONS

'Carve an indentation in a brick, fill it with crushed basil, and cover the brick with another, so that the indentation is completely sealed. Expose the two bricks to sunlight, and you will find that within a few days, fumes from the basil, acting as a leavening agent, will have transformed the vegetable matter into veritable scorpions . . .

'If a soiled shirt is placed in the opening of a vessel containing grains of wheat, the reaction of the leaven in the shirt with fumes from the wheat will, after approximately twenty-one days, transform the wheat into mice.'

Jan Baptiste van Helmont

directly to frogs, ants and leeches as well as plant material. He left instructions for producing scorpions from basil leaves and sunlight, and mice from a dirty shirt and a jar of wheat (see box above).

Testing times

Not everyone was persuaded that animals could spring from inert material. In 1668, the Italian biologist Francesco Redi carried out a simple experiment to test the idea that maggots generate spontaneously in rotting meat.

Three pieces of meat were put into three separate jars. He left one open to the air, covered the second one with netting and sealed the third one entirely. As we would (now) expect, the sealed jar didn't grow any maggots, but the meat in the open jar was soon crawling with them. Most revealingly, maggots appeared on the inside of the net covering the second jar. Flies had laid their eggs through the holes in the net, so that was where the maggots hatched. Redi's results persuaded many scientists of his day that maggots, at least, are not produced by spontaneous generation but from very small eggs laid by flies.

Even so, a century later, the French scientist Georges-Louis Leclerc, Comte de Buffon, was still confident that at least the smallest organisms are produced only by spontaneous generation. He wrote in 1777 of the way chemicals are freed in decaying meat (or bodies):

'Organic molecules roam freely within the matter of dead, decomposed bodies . . . these organic molecules, always active, rework the putrefied substance, appropriating coarser particles, reuniting them, and fashioning a multitude of small organized bodies. Of these, a few, like earthworms and mushrooms, resemble relatively large animals or vegetables, while the others, almost infinite in number, are visible only under a microscope. All such bodies come about only by spontaneous generation.'

By this time, Pier Antonio Micheli had already discovered that he could grow fungus by placing spores from a fungus on slices of melon; the fungus that grew was of the same type as the source of the spores, leading him to conclude that fungi are not produced by spontaneous generation.

Boiled broth

The Italian biologist Lazzaro Spallanzani carried out a wider-ranging experiment in 1768. He poured broth into flasks, expelled enough air to prevent the flasks from exploding, sealed them and boiled the broth. As he expected, the broth remained good as long as he left the flasks sealed. This didn't persuade everyone, though: it was easy to argue that by excluding the air he had suffocated the microorganisms or prevented them from generating.

The French biologist Louis Pasteur was more rigorous and it was he who dealt the death blow to spontaneous generation. He boiled broth in two flasks with curved swan-necks. The flasks were not

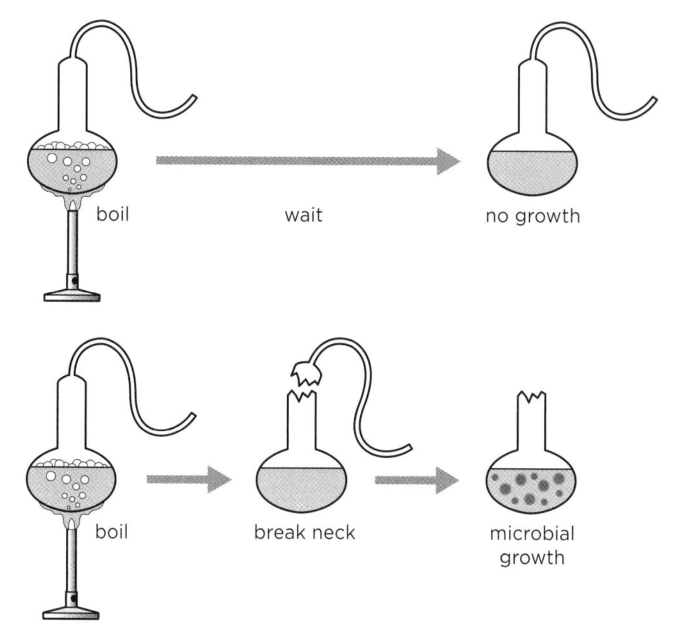

sealed, so he could not be suffocating any nascent microorganisms. After boiling, he snapped the neck off one flask but left the other intact. The broth in the broken flask soon began to spoil and Pasteur found microbes in it, whereas the broth in the intact flask did not spoil and there were no microbes. Although air could enter the flask, microbes could not: they could enter the open neck of the flask but could not travel up the S-bend of the flask. The microbes that went in simply piled up at the bottom of the bend.

Pasteur's experiment should have ended the spontaneous generation debate once and for all, but a few old stalwarts carried on defending it, against all reason, in subsequent decades.

> ***'Never will the doctrine of spontaneous generation recover from the mortal blow of this simple experiment. There is no known circumstance in which it can be confirmed that microscopic beings came into the world without germs, without parents similar to themselves.'***
> Louis Pasteur

And yet . . .

The origins of life remain obscure. Theories about the first life on Earth range from the belief that it arrived from elsewhere in space (which just moves the question further along) to various ideas about how chemicals might combine together to form compounds that can make copies of themselves and develop eventually into something we could call life (see page 262). It doesn't help that we can't quite define life. But at some point, unless we resort to a supernatural entity, there has to be a stage when life is 'turned on', animating previously non-living components. If life emerges when the right combination of chemicals is put together in the right order it still, originally, generated spontaneously. The best we can say is that current life-forms do not generate spontaneously, but all come from pre-existing life by natural processes of reproduction.

THE BOUNDARIES OF LIFE

Scientists disagree about whether viruses should be considered alive. Viruses can't reproduce on their own – they need to co-opt matter from a living host cell to make copies of themselves. But they can reproduce, given these conditions, and they do have genetic material which is passed on from one generation to another. Their genetic material can change, by mutation, producing new strains.

It might not seem as though it matters very much whether viruses count as living things – but if they do, they are by far the most numerous life forms on Earth.

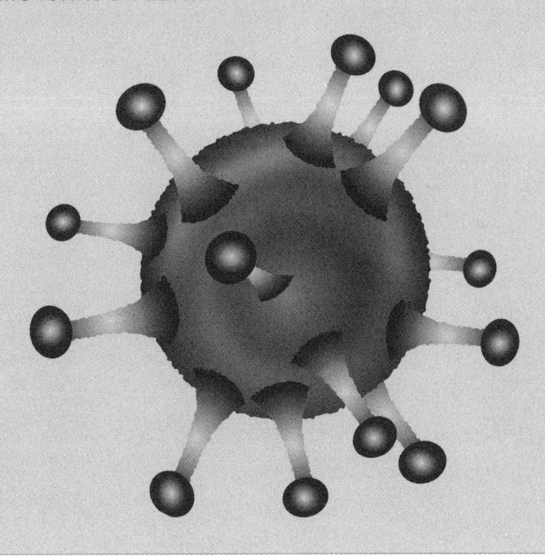

Chapter 3
Attachment Theory

Does lots of affection and attention spoil a child?
Or is it essential to healthy development?

THE IDEA

A strong attachment (emotional bond) to at least one person is essential to a child's healthy emotional, psychological, social and perhaps physical development.

Seen and not heard?

Until the early years of the 20th century, children in Britain and North America often suffered under a harsh regime. They were considered best seen and not heard (and possibly not even seen). Even by the middle of the century, in wealthier households they might spend more time with a nanny or governess than with a parent. Some were sent away to boarding schools at a young age, and scant open affection was shown to them at home. This behaviour, which now seems neglectful and brutal to many people, was supported by contemporary theories about children and child-rearing.

It was commonly believed that small children were attached to their parents simply because their parents provided them with food. It was thought that they were more likely to be attached to their mother, as the mother had the milk supply. As far as there was any psychological basis for the belief, it was that the infant came to associate the mother with feeding through a process of classical conditioning (see box opposite). The child had learned to associate the mother with milk; milk was a good thing, so the mother was a good thing.

> ***'Never hug and kiss them, never let them sit in your lap. If you must, kiss them once on the forehead when they say goodnight. Shake hands with them in the morning . . . When you are tempted to pet your child remember that mother love is a dangerous instrument. An instrument which may inflict a never-healing wound, a wound which may make infancy unhappy, adolescence a nightmare, an instrument which may wreck your adult son or daughter's vocational future and their chances for marital happiness.'***
>
> John B. Watson

Another way of thinking

The Behaviourist approach dominated psychology in the early 20th century, but one man, John Bowlby, turned against it. Bowlby was greatly influenced by the work of a zoologist, Konrad Lorenz, who

DRIBBLING DOGS AND CUPBOARD LOVE

In the 1890s, the Russian physiologist Ivan Pavlov was working with dogs. He discovered accidentally that the dogs' natural inclination to salivate (drool) when they saw, smelled or tasted food could be transferred to something they associated with it – such as himself. Because he was the one who fed the dogs, they salivated when he entered the room, even if he didn't have any food with him. He decided to investigate this learning mechanism and discovered the process now called classical (or Pavlovian) conditioning. The dogs were taught to associate an unrelated stimulus, such as hearing a bell or whistle, with receiving food. Once the dogs learned the link, their involuntary response – salivating – was produced when they heard the sound.

Pavlov published his findings in 1902 and they had a great impact on theories about learning and behaviour. An entire psychological movement, Behaviourism, developed from the belief that all behaviour is the result of learned responses. This suggests that people are essentially programmed by their early experiences and education. Babies, then, are programmed to like their mothers and want to be near them because the mother provides milk.

MOTHER GOOSE

The Austrian zoologist Konrad Lorenz took a clutch of goose eggs and split it into two groups. He allowed the mother goose to incubate and hatch one set, and he hatched the other set in an incubator, making sure he was the first living being those goslings saw. The goslings 'imprinted' on him, treating him as their parent and following him everywhere. They showed no recognition of their real mother.

Lorenz's experiments showed that imprinting usually happens 12 to 17 hours after hatching. Importantly, he found if goslings saw nothing to imprint on within 32 hours it was too late – the critical period had passed and they would not imprint. He also found imprinting to be irreversible. Once goslings had imprinted on him, they would not change their minds and prefer a real goose. They would continue to follow him around as they grew up.

worked with newly hatched birds (see box opposite). Lorenz had discovered that goose chicks become attached to their mother without receiving food from her. Indeed, he showed that attachment, at least in chicks, is innate – chicks seem to be programmed by evolution to find and follow their mother, without needing to learn anything.

Two models – innate or learned?

Lorenz's results suggested that the connection between mother and infant is innate – a natural, automatic connection that happens without the infant needing to spend time learning. This was the opposite of the Behaviourist model, which required the infant to learn the association between the mother and food in order to develop an attachment.

Children need to be loved

In the 1930s, Bowlby worked in a clinic where he dealt with emotionally disturbed children. He noticed that separating an infant or child from his or her mother caused distress, regardless of whether

the child was deprived of material things the mother generally supplied: even a well-fed child suffered from separation. Then he carried out experiments in which he separated small children from their primary caregivers for short and longer periods and found a pattern in their responses. They would first cry, then calm a bit, but remain withdrawn and unresponsive. After still more time, the children began to interact with others, but were angry at the caregiver on their return. He described a pattern of protest, then despair and finally detachment.

Bowlby concluded that the infant forms a close emotional bond with a primary caregiver, usually the mother, and that any threat to this bond causes distress, even if the child is adequately fed by someone else. He suggested that evolution has equipped children and parents with this bond as a protective mechanism.

He explained the evolutionary strategy that makes mother love far from a 'dangerous instrument'. The child has developed actions (called 'social releasers') such as smiling and crying that a parent responds to with attention and care. Smiling and crying are innate; the child does them automatically in appropriate circumstances.

The child responds to the parent's response by growing attached and trusting the parent, and the child's trust and affection reinforce the parent's behaviour and attachment to the child. Protection and help from the parent improve the child's prospects of survival, so attachment is reinforced by evolution: children prone to form a strong attachment will survive and reproduce, passing on the genes for forming attachment.

Bowlby defined attachment as 'a lasting psychological connectedness between human beings' and declared it essential to healthy development. Without this bond forming with at least one person in the first couple of years of life, he proposed, the child is at risk of emotional disturbance, physical and psychological ill-health and problems with development.

Orphans and hospital visits

Bowlby's work was revolutionary. He advocated emotional engagement with children, spending time with them, showing them affection and being supportive and encouraging – the exact opposite of the advice of the Behaviourist school.

It had an impact on practice almost immediately. The World Health Organization (WHO) worked with Bowlby to develop strategies for dealing with orphans after World War II. In 1952, Bowlby made a short film called *A Two-Year-Old Goes to Hospital* about the

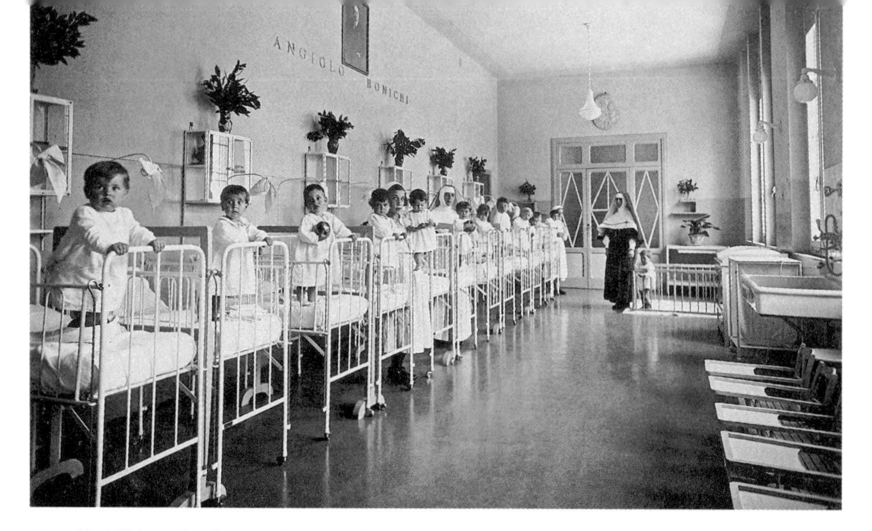

Small children in the regimented dormitory of an orphanage in Italy during World War II.

distress suffered by young children separated from their parents when they go to hospital. As a result, hospitals began to recognize that children's misery was more than just fussing and started to encourage parental visits and involvement.

Watching attachments grow

In 1964, two psychologists, Rudolph Schaffer and Peggy Emerson, followed the development of a group of infants over 60 months, tracking attachment as it emerged. They found that babies' development can be divided into stages:

FORTY-FOUR THIEVES

Bowlby carried out a study in the 1930s with 44 young offenders convicted of theft, which he cited in support of his theory. Comparing them with a control group of 44 adolescents who had not been convicted of a crime, he found:

- A disproportionately high number of young offenders (more than half) had suffered a period of separation from their primary caregiver of at least six months in the first five years of life; only two of the control group had been separated from their caregiver.
- A third of the offenders displayed 'affectionless psychopathy' – they were unable to show affection or concern for others and acted entirely selfishly. None of the control group did.
- Of the 14 'affectionless psychopaths', 12 (86 per cent) had endured a period of at least six months' separation from their caregiver in early life.

Bowlby concluded that separation from the attached caregiver (usually the mother) led directly to delinquent behaviour and to stunted emotional development.

Critics of his study have pointed out that he found a correlation rather than causality; other factors could have been involved and it was not reasonable to assume that separation directly caused criminal behaviour. For example, children separated from a caregiver might have been more likely to grow up in poverty, or have a poor diet, which could have led to their poor outcomes.

HANDS-OFF PARENTING

Bowlby's own upbringing probably influenced his research interests. His mother was distant and saw him for only an hour a day after teatime. He was raised primarily by a kindly nanny, whom he loved, but she left the family when he was four years old, to be replaced by someone with a much colder attitude. When his father went to serve in World War I, Bowlby had no contact with him. Bowlby was seven years old at the start of the war, and was sent to boarding school at the age of ten.

0–6 weeks, an asocial stage: the baby smiles at many types of stimulus, whether or not it involves engaging with other people.

6–42 weeks, indiscriminate attachments: the child will engage with anyone who engages with them, becoming upset when a person stops engaging. From around 3 months, they start to show a preference for familiar caregivers, and those people can calm or comfort the child more easily than strangers.

7–9 months, specific attachment: children show a preference for a single caregiver and look to familiar people for comfort and security. They show anxiety when separated from their special person, and are wary or afraid of strangers.

10+ months, multiple attachments: most children form a specific attachment by 12 months. The child also forms secure attachments with other people (for example, another parent, siblings, grandparents).

Schaffer and Emerson found that the most important factor in the child forming an attachment was responsiveness from the adult. Interacting and taking notice of a child, playing with the child, responding quickly and offering comfort and engagement were more important than being the one who provides food and practical care.

Cloth and wire mothers

The distinction between affection and practical care was explored by the American psychiatrist Harry Harlow in the 1950s and 1960s. He carried out a series of brutal experiments with baby rhesus monkeys to explore the nature of attachment. The experiments have since been severely criticized on ethical grounds, as they caused extreme suffering and irreparable psychological harm to the monkeys involved.

Harlow reared baby monkeys in isolation for periods of three, six, nine and 12 months. He then reintroduced them to other monkeys and observed their behaviour. At first they were afraid of the others and then aggressive towards them. The isolated monkeys couldn't communicate or socialize, were soon bullied and began self-harming. How badly they were affected was directly related to how long they had been isolated, with those kept apart for 12 months never recovering enough to socialize with other monkeys.

A baby rhesus monkey would visit the wire 'mother' only for food and then retreat from its comfortless companion.

He carried out other experiments in which infants had access to two surrogate model mothers. One model mother was a bare metal framework and the other had a cloth covering. He found that infants sought comfort by snuggling against the cloth mother. Then he installed bottles so that some of the model mothers provided milk.

If milk was provided by the metal surrogate, the infants would go to it just to feed and then return to the cloth mother. They fared better than infants who were entirely isolated, but still had development problems: they were insecure and timid, could not socialize with other monkeys, were bullied by other monkeys, had difficulty mating and the females became inadequate mothers.

The infants reunited with real monkeys before 90 days recovered and formed attachments – so 90 days was the critical period in which an attachment must be formed in rhesus monkeys.

Harlow also found that infant monkeys brought up in isolation but allowed 20 minutes a day in a playroom with three other monkeys developed to be emotionally and socially normal.

Harlow's discovery that infant monkeys spent more time with the cloth mother (which provided physical comfort) than the metal mother (which provided food) supported Bowlby's view that

PRIVATION AND DEPRIVATION

Two different types of separation from an attachment figure have been studied: privation and deprivation. Privation occurs when the infant has no attachment to an adult; deprivation occurs when an attachment is developed but then broken. It can result when a parent dies or falls seriously ill, for instance. The effects of privation are more severe than the effects of deprivation.

attachment is related to responsiveness and comfort rather than to food. The sad fate of the isolated infant monkeys supported his theory that attachment is essential to mental health in later life.

Not just monkeys

In 1989, the deplorable state of children kept in orphanages in Romania was revealed to the world. Many were tied to cots or beds, had no physical contact with a caring adult and were emotionally and physically deprived. Some of the children were rescued and put with foster families and their development was tracked over the coming years. The children of the Romanian orphanages have been the subject of many studies in psychology and have provided much of the solid information we have about the impact of parental care and attachment in humans.

Many of the orphans never recovered, and were both physically and emotionally damaged. Brain scans of children who had spent years in the orphanages found that their brains were smaller than average for their age and had lower levels of activity than normal. A functional brain scan, watching the brain while it is working, showed that the part of the brain that is active when someone responds emotionally did not distinguish between the child seeing their foster parent and seeing a stranger. The children who left the orphanage by the age of two generally made a full recovery, but some of those who stayed longer were permanently damaged. (The children in the orphanages also suffered from malnutrition, lack of exercise and other types of hardship, so the correlations are not entirely straightforward.)

Patterns for the future

Bowlby claimed that among the advantages of forming a healthy and strong attachment as a small child is the ability to grow up to be happy, confident, socialized and intelligent. The relationship with the attachment figure gives the child a template for forming relationships in the future and helps to build their internal working model - a framework for understanding the world, themselves and other people. If the relationship is strong and healthy, the child will be secure and confident.

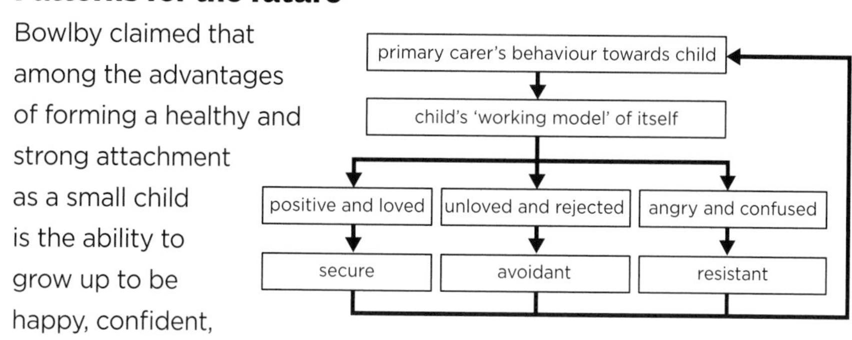

There is no corresponding body of evidence to support Watson's belief that 'mother love is a dangerous instrument . . . which may inflict a never-healing wound'. If you're looking for a theoretical basis for your approach to child-rearing, attachment theory is a good start.

Chapter 4
Theory of Evolution

Why can't penguins fly? Why are wolves grey?
Why are living things the way they are?

Bare bones

The theory of evolution explains why life on Earth is diverse and how species change over time. It's most often associated with the 19th-century naturalist Charles Darwin (see page 54), but has itself evolved over the last 150 years. In its current form, called the modern evolutionary synthesis, it is combined with the theory of genetic inheritance (see page 234) to offer a near-complete explanation for the development of life on Earth. The scientific community accepts evolution as the best account of the development and extinction of species, but some religious groups reject it, favouring instead a belief in divine origins for life. Creationism is not a competing theory; it is a belief accepted without critical, scientific scrutiny.

THE IDEA

Organisms change over time as a result of natural selection; those best suited to survive in their environment and prevailing conditions are most likely to succeed and reproduce, so their characteristics are passed on to future generations. Those less well adapted die without reproducing.

CHARLES DARWIN, 1809–82

The eminent naturalist Charles Darwin published his theory of evolution in *On the Origin of Species by Means of Natural Selection* in 1859. He developed it from his observations of wildlife and geology on a round-the-world trip on HMS *Beagle* in 1831–6, followed by 20 years spent refining his ideas. During the trip, he encountered plants and animals that had developed independently in different areas, but had often developed the same type of solution to similar problems. This led him to theorize that organisms have diversified from a few common ancestors in a process that has taken millions of years and still continues. When Darwin first proposed the theory it was met with opposition and ridicule from many quarters. His contemporaries were resistant to the idea that humans don't hold a special place in the natural world but are animals that have evolved from earlier species.

Ideas about origins

Darwin was not the first person to suggest that organisms had changed over time. As early as the 6th century BC, the Greek philosopher Anaximander posited that the very first creatures had emerged from bubbling mud in the early days of the Earth. As land and water separated, some of these creatures adapted to life on land and some to aquatic life. Even humans, in his view, had developed from these proto-fish. Around 50 years later, Xenophanes recognized fossils as relics of long-dead life forms and proposed that the land and sea had changed places in the past and would do so again.

Yet this insight was lost for millennia. A competing idea, that the Earth had been created in its final form, was supported by religion and became the prevailing model. The story of Creation in Genesis and in the Quran had God or Allah creating all of the creatures and humankind at a stroke. There was no need, and no excuse, for suggesting that things changed. It was only in the 18th century that the idea of developing organisms arose again.

Time for change

Among the first people to propose something like evolution was the French theoretical scientist Pierre de Maupertuis. He argued that organisms take features from each of their parents, but new or different features might sometimes arise by chance and these can be maintained if successful. He did not rule out the possibility of the environment having an impact on organisms, producing changes in them that could be passed on to their offspring.

> ***'Do not be angry if I say you were a worm, or an egg, or even a kind of mud.'***
> Pierre de Maupertuis, *The Earthly Venus* (1745)

Georges-Louis Leclerc was also influenced by de Maupertuis. He noticed that very different animals lived in the forests of South America, Asia and Africa, even though the climate was similar in all three regions. This led him to suggest that God had created all of the animals near the North Pole (at a time when it was warmer there)

> ***'Would it be too bold to imagine, that in the great length of time, since the earth began to exist, perhaps millions of ages before the commencement of the history of mankind . . . that all warm-blooded animals have arisen from one living filament, which THE GREAT FIRST CAUSE endued with animality, with the power of acquiring new parts, attended with new propensities, directed by irritations, sensations, volitions, and associations; and thus possessing the faculty of continuing to improve by its own inherent activity, and of delivering down those improvements by generation to its posterity, world without end.'***
> Erasmus Darwin, *Zoönomia* (1794–6)

and as they spread out they had adapted to the habitats in which they found themselves. He further proposed that humans and apes might have a common ancestor – enough to arouse the suspicion of the Church at the time.

The first hint of evolution in the modern sense came from Darwin's own grandfather, Erasmus Darwin. He wrote in *Zoönomia* of all life coming from a single common ancestor, forming 'one living filament'. He suggested that the strongest and 'fittest' organisms (those best suited to their environments) are those which get the chance to reproduce, so their characteristics are passed on.

Improving through effort

Soon after, the French naturalist Jean-Baptiste Lamarck proposed the first proper evolutionary theory. He suggested a gradual, smooth change over time with the 'life force' driving evolution to produce ever-better organisms. Evolution was thus purpose-driven, progressive and cumulative, leading to the improvement of all species. Later organisms were better than earlier ones as they had been refined.

Further, Lamarck saw evolution as the result of striving on the part of previous generations. According to his model, the giraffe has evolved a long neck because countless generations of giraffes have stretched to reach high leaves. Each one that stretched its neck a little more passed on a slightly longer neck to its offspring.

Darwin's theory

The theory of evolution propounded by Darwin didn't have organisms evolving towards any goal, or even necessarily improving over time. He saw instead a steady process of adaptation. The full title of the book in which he explained his theory is *On the Origin of Species by Means of Natural Selection*. By natural selection he meant that the organisms best suited to survival in their environment are those able to reproduce and pass on their genetic inheritance.

Darwin thought that variation arises naturally and by chance in organisms; some new features will be advantageous and some disadvantageous. Those that are advantageous will be preserved because the individuals best suited to live in prevailing conditions will survive and breed. Because organisms are always in competition with one another – for food, living space and mates – those that are successful will be those best adapted, and over time the features that make them best adapted will come to predominate. Eventually, they can become defining characteristics of the species.

As organisms spread out into different environments, or conditions change, different features might be needed to remain successful. This can lead to diversification and eventually the development of new species as they adapt in different ways.

'Nothing in biology makes sense except in the light of evolution.'
Theodosius Dobzhansky, 1973

Darwin's evidence

Among the evidence Darwin drew on to support his theory was the variation he found in the finches living on the Galapagos Islands. This group of islands off the coast of Ecuador is home to several species of finch (a type of small bird) which look alike in many ways but have markedly different beaks.

Darwin saw sufficient similarities between the finches to conclude that they all shared a common ancestor – that is, they all developed originally from one type of bird. He assumed the ancestors of the finches had all originated from the mainland (Ecuador), but the populations had become isolated on the individual islands.

The islands afforded different types of food. On each island, the successful finches were those with beaks adapted to eat food that was readily available and not sought by numerous competitors. Over generations, the beak shapes became increasingly refined and specialized so that each type of finch was highly adapted to its particular environment.

Successful and sexy

Darwin saw natural selection working in two ways. One, later dubbed 'survival of the fittest', meant the best-adapted organisms were most likely to survive and reproduce. The other was sexual selection. This meant that some individuals were more likely to reproduce than others because they were more attractive to the opposite sex. In some cases this might reinforce selection by fitness: if an animal chooses a mate good at finding food, or that is larger and stronger than others, for instance, this selects for survival traits.

In other cases, sexual selection is not related to survival. For example, female peacocks select male peacocks with the most decorative 'eyes' on their tails, and with the largest tails. Over time, sexual selection by females leads to peacock tails becoming larger and more decorative, but this does not make the peacocks better at finding food or avoiding predators. Indeed, a large and noticeable tail might even be a disadvantage, making it harder for the peacocks to walk through undergrowth or flee from predators.

The modern evolutionary synthesis

Darwin had no idea how the characteristics of an organism were passed to offspring and down the generations. At the time he was writing, the pattern of inheritance had not yet been publicized, and the discovery of genes and DNA (deoxyribonucleic acid) lay far in the future (see page 234). But as these were uncovered, the mechanism lying behind evolution became clear.

The pieces were put together by Theodosius Dobzhansky in 1937. In *Genetics and the Origin of Species*, he defined evolution as 'a change in the frequency of an allele within a gene pool'. (An allele is one of two or more possible genes for a specific feature, such as brown fur.) He saw natural selection progressing by a process of mutation, with the most beneficial mutations being favoured by competition. Mutation is not targeted, nor does it tend to produce improvements. It is random and often produces damaging or unhelpful variations that make no difference to the organism's prospects for survival.

Unfavourable mutations will disappear rapidly, as the individuals displaying them are less likely to survive and reproduce. Variations which make no difference to the organism's survival prospects might coincidentally become more common in a species, finally changing its character in a process called 'genetic drift'. An example would be if a plant that usually had yellow flowers developed a variant with white flowers, which was equally good at attracting pollinators. Random chance could lead to the white form becoming predominant over time. Genetic drift happens most frequently in small, isolated populations.

Slowly does it?

Darwin believed that evolution progressed slowly and steadily, but this need not always be the case. In 1972, evolutionary biologists Stephen Jay Gould and Niles Eldredge suggested the possibility of 'punctuated equilibrium', or short bursts of evolutionary activity happening after long periods of slow or no change. Their work showed that there are often long periods when there is little change in the fossil record, and then a period of rapid change. This can happen in the evolution of individual species or after catastrophic mass extinction.

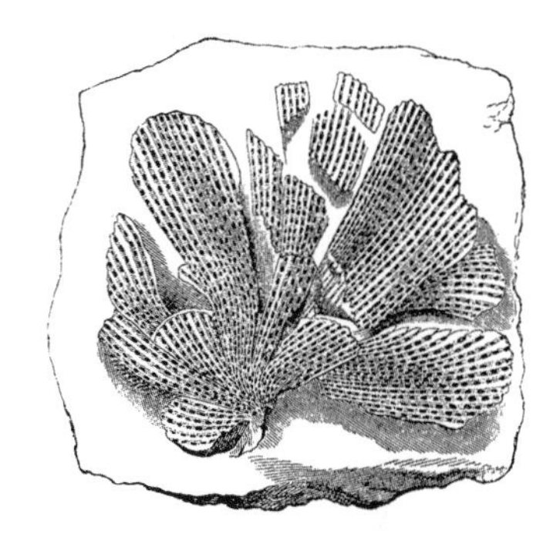

In the case of a single species, rapid change tends to happen at the edge of the geographical area occupied by a population, or in isolated groups. There is fossil evidence for punctuated equilibrium in some types of bryozoan (a coral-like sea creature). On the whole, the fossil record preserves so few of the individuals which have ever lived that periods of rapid change are hard to find.

Selfishness and altruism

Darwin showed individual organisms competing against one another, but also noted that some animals act with altruism, putting the benefit of another ahead of their own. He suggested that altruism might promote the survival of the whole group, and this notion remained popular until the 1960s. But it's hard to see how evolution can work at the group or species level. An alternative is to take a step down from the organism and look at individual genes as the drivers of evolution. Richard Dawkins took this path in his book *The Selfish Gene*, published in 1976.

Dawkins suggested that all acts which look like altruism are the result of selection at the level of genes. A parent of any species caring for its offspring seems altruistic: the parent gains nothing from the time invested in looking after, say, the baby pangolin. The parent could be out eating all the worms itself, or spending its money on holidays and fast cars, depending on its species. If all individuals did this, though, the species would die out very quickly. There is a biological imperative to reproduce and care for young because those individuals which didn't have genes prompting them to reproduce have not passed on their genes. It's quite possible that the pleasure we gain from our children is the genetic reinforcer which leads us to care for them.

But parent–child altruism is only part of the deal. Imagine an animal that spends time grooming others in its group but does not eat the pests it picks out. This looks altruistic, but animals that groom one another generally do so because they are more

likely to receive grooming in return. There will be a few freeloaders, who accept grooming but never reciprocate. But their genes never come to predominate in a population, because if there are too many freeloaders it becomes harder to find a groomer, and freeloaders start to suffer from the pests that bother the ungroomed. The grooming gene will predominate in a group because it leads to survival. Maths will take care of the ratio of grooming and non-grooming individuals in a population.

Even behaviours which look entirely altruistic can have a selfish motive. Among some animals living in social groups, such as meerkats, individuals that sense danger will alert others in the group, making themselves more visible and vulnerable as a result. It might seem that they would do better to run away and leave someone else to be eaten. But usually animals that issue such warnings prosper – and perhaps only survive – in large groups. There is no point, in terms of the future of its genes, in an animal saving itself if it is then doomed because it has no social group in which to live and reproduce. It's easy to think of something like this sort of sentinel behaviour as evolution acting at the level of

Dawkins suggests that the tales of dolphins magnanimously saving divers and sailors are actually cases of mistaken identity; the dolphins probably mistake the humans for distressed dolphins, to which they might be related.

the group, because we see the effect in terms of the group. But this effect emerges from the combination of individuals behaving in their own interest, and is the mathematical result of gene selection.

From tree to cladogram

The only illustration in *Origin of Species* is a 'tree of life' – a diagram like a tree showing proposed evolutionary relationships between different types of organism. The tree of life replaced the earlier 'chain of being' or 'ladder of nature' – both old models that placed organisms on a hierarchy of creation from the 'lowest' forms such as fungi to the 'highest' form: humans. While the tree structure is less hierarchical and aims to show how organisms are similar and where they diverge from one another it has now been replaced by an even less hierarchical image, the cladogram. This shows precisely

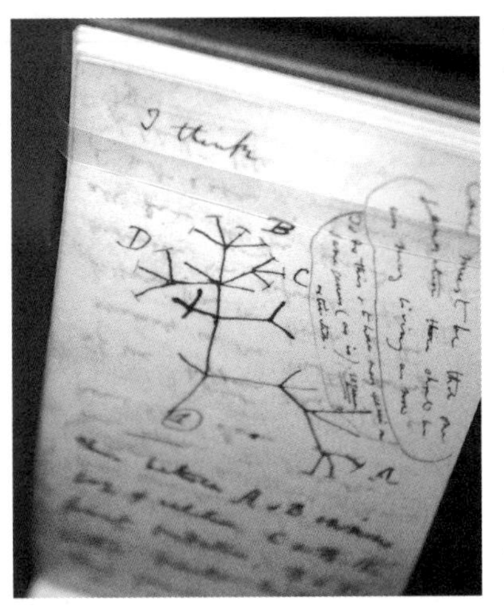

A 'tree of life' sketch in Darwin's original notebook.

where organisms branch off from a hereditary line by developing a significant difference.

The original method of classifying organisms relied on looking at their body structure and appearance (morphology) and how their bodies work (physiology). It could easily lead to mistakes. Is a bat a bird? Is a dolphin a fish? They both have features that could tempt us to put them in the wrong group. Today, gene sequencing technology has made defining genetic relationships between species or even individuals much more precise. It means we can state the differences between species objectively and assign individuals to a species with greater accuracy.

The most important impact of gene sequencing, though, has been to show that the theory of evolution is essentially correct. We can find common portions of DNA in widely differing organisms and trace chunks of DNA through the evolutionary tree or cladogram. The evidence for diversification as species split is preserved in the DNA which organisms share.

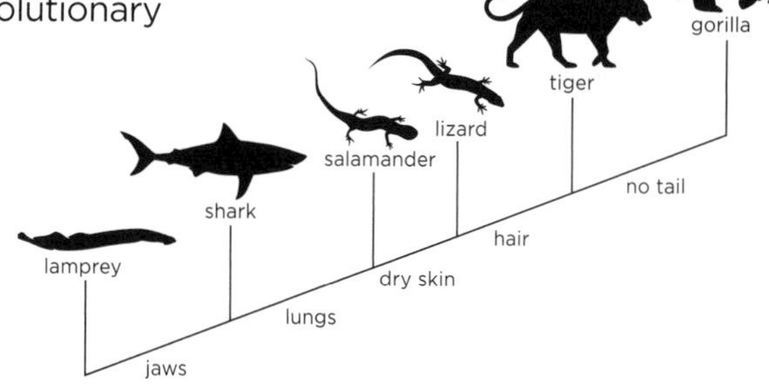

Chapter 5 Climate Change Theory

There is little doubt that the climate is changing – but is change caused by human activity?

All change?

The climate on Earth is a product of many factors working together, including the composition of the atmosphere, the activity of the Sun, minor changes in Earth's orbit around the Sun and the location and form of landmasses on Earth.

The aspects of climate that most interest humans are the temperature and the prevalence of dangerous weather events such as storms and hurricanes. The climate is undeniably changing – but it is always changing. What is remarkable now is how quickly it is changing compared to what we know of the patterns of climate change in the past. The most important questions for humans

Human activity is increasing the proportion of carbon dioxide (CO_2) and other greenhouse gases in Earth's atmosphere, leading to the planet warming and the climate changing. This is also known as Anthropogenic Global Warming (AGW) theory.

CLIMATE CHANGE V. GLOBAL WARMING

Most climate scientists prefer the phrase 'climate change' to 'global warming' for several reasons. Climate change is about more than temperature. It includes patterns of rainfall, prevalence and ferocity of storms and wind, and changes to currents in the air and consequently the sea.

One effect of current climate change is rising global temperatures, which is where the label 'global warming' comes from. The global temperature is the average temperature over the whole world. Global sea temperature is measured separately. Changes in the average temperatures of air and sea bring about changes in weather patterns.

are whether we are causing it to change, how the changes will affect us and whether we can do anything to prevent, slow or counteract the likely effects of climate change.

Dinosaurs in the swamp

Climate scientists are able to work out what the climate has been like in the past by looking at geological records. It's clear that the temperature and the atmospheric composition on Earth have been very different in the past. For example, 500 million years ago, when animals first crawled from the water and started to explore the land, the world was about 15°Celsius hotter than it is now. Two hundred million years later it plunged to 2°C below current levels and 50 million years after that it was back up to 11° above current levels. The last major peak, at +14° compared to now, was 50 million years ago. Since then, the trend has been generally downwards, with occasional short-lived peaks, until 20,000 years ago. It has been fairly stable at the current level for around 10,000 years – the whole period of human civilization.

There is no question that Earth can sustain life at much higher – and probably lower – temperatures than at present. But there is no reason to suppose human life can be sustained at vastly different temperatures. Average temperatures have changed within small margins since the start of civilization. A warmer period in the Middle Ages saw grapes growing in England and a cooler period in the 17th and 18th centuries saw rivers in Europe freezing (they seldom freeze today). The cold period caused many deaths and during the warmer periods some areas that are now inhabited were under water. Significant changes in temperature now would certainly have an impact on our lives.

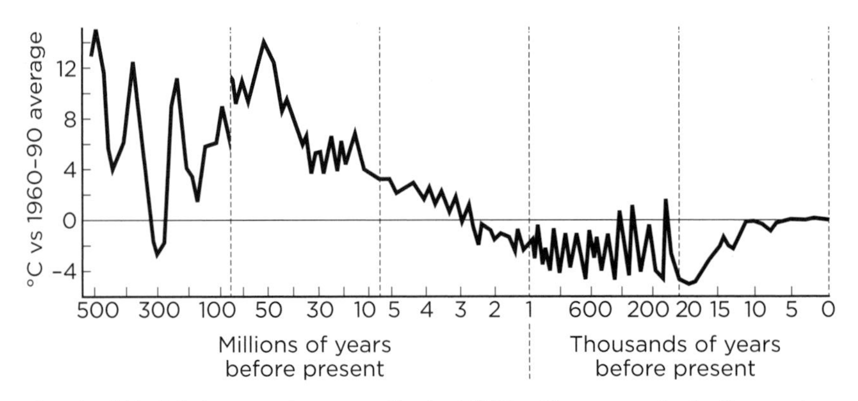

Graph of Earth's temperature over the last 500 million years (note the x-axis changes scale several times).

Why does the climate change?

The climate changes for many reasons. The distribution of land masses has an impact, but this change takes place very slowly as the continents move (see page 142). Solar activity seems to have affected the climate on Earth at some points in the past, but there has been no significant change in patterns of solar activity since 1978, while the temperature on Earth has been rising. Slight changes in the Earth's orbit around the Sun can lead to hotter summers and cooler winters, but these orbit changes happen gradually, over thousands of years. Whether land is covered with forest, is bare or is built on alters how air flows over it and affects patterns of rainfall and temperature.

> **CLIMATE CHANGE OR ODD WEATHER?**
>
> Climate is not the same thing as weather. Weather describes atmospheric conditions in the short term – whether it is hot or cold today, or raining, snowy or foggy, and so forth. Climate relates to long-term patterns in the behaviour of the atmosphere.
>
> It's easy to think of individual weather events or hot or cold years as supporting or refuting climate change theory, but short periods and single events are not significant. Trends are important. One warm winter doesn't count for anything – but a trend of five unusually warm winters out of six begins to look significant.

The composition of the atmosphere, including the proportion of carbon dioxide and some other gases, might have a rapid impact on the climate. Carbon dioxide is removed from the atmosphere by plants photosynthesizing and is added to the atmosphere by volcanic activity, breathing animals (and respiring

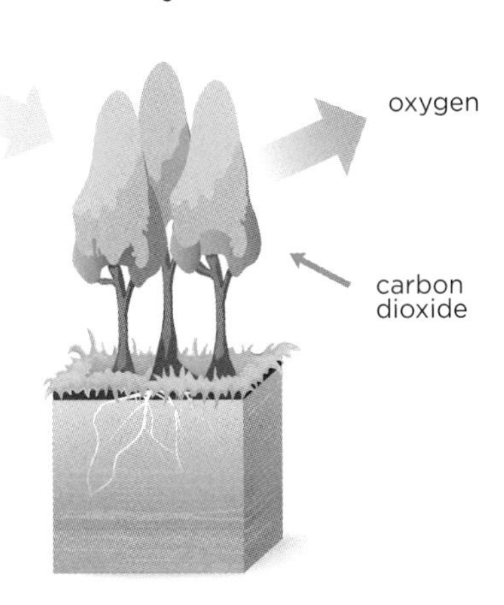

plants) and by burning fossil fuels and wood. Carbon dioxide can also be dissolved in the sea and released according to the temperature of the water.

> ***'Observations throughout the world make it clear that climate change is occurring, and rigorous scientific research demonstrates that the greenhouse gases emitted by human activities are the primary driver.'***
> Statement on climate change from 18 scientific associations, 2009

Average temperatures on land and sea are rising. More than 97 per cent of active climate scientists agree that human activity is responsible for at least some of the change, pointing to our use of fossil fuels, land clearance and farming and industrial practices. Climate change theory proposes that all these lead to changes in the atmosphere, increasing the concentration of gases that lock heat in and thus producing the rise in temperature.

In the greenhouse

There are several gases which fall under the term 'greenhouse gas'. The most important are water vapour, carbon dioxide, methane and nitrous oxide. Greenhouse gases form a layer high in the atmosphere which acts a bit like an insulating blanket. As heat from the Sun strikes the Earth (as infrared radiation), some is reflected back into space and some passes through the atmosphere to warm the ground and the sea. Heat is released from the ground again as infrared radiation. Some of this escapes into space and some is trapped by greenhouse gases and warms the atmosphere.

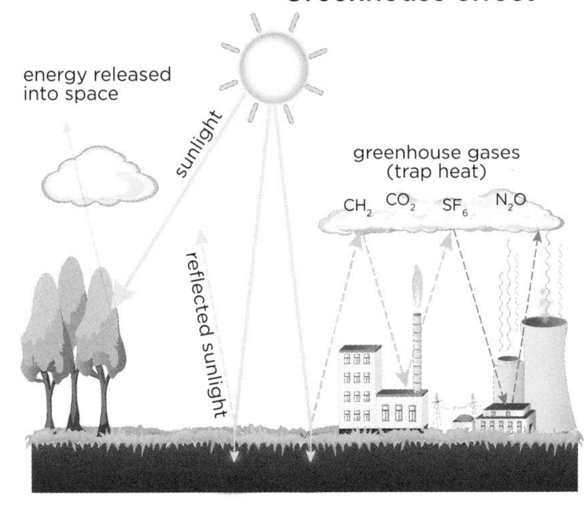

Not all greenhouse gases are equal. Some trap more heat than others and some hang around in the atmosphere for longer than others. Every kilogram of methane traps 21 times as much heat as every kilogram of CO_2, so although we produce less methane than CO_2, what we do produce has a big effect.

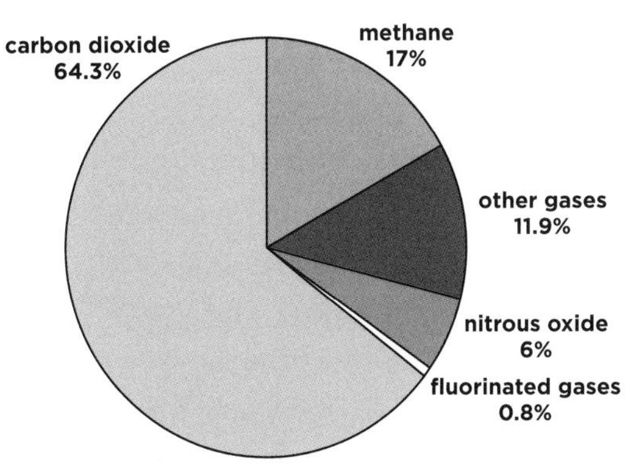

If a planet has too high a proportion of greenhouse gases, it falls into a runaway heating cycle. That has happened on Venus, which has a surface temperature of more than 460° C. If there is too low a proportion of greenhouse gases, a planet's temperature see-saws wildly between day and night as it can't keep its heat when not directly warmed by the Sun – infrared radiation escapes straight into space. For a planet to be hospitable, it needs just the right amount of greenhouse gases.

For life-forms that have evolved to suit a particular pattern of temperature, changes in the amount of greenhouse gases, and subsequently of temperature, can be devastating. Change doesn't mean the end of life on a planet, but it can lead to mass extinction and the evolution of new life-forms better adapted to the new pattern. Climate change is thought to have been wholly or partly responsible for some of the mass extinction events in the past.

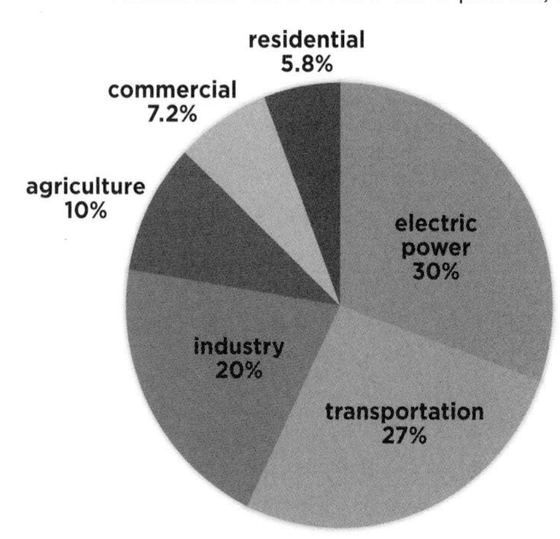

US greenhouse gas emissions by sector of the economy, 2014.

Spreading it about

Human activity adds greenhouse gases in various ways. Most of it comes from burning fossil fuels to generate energy, whether for heating and lighting at home, or for industry, or for transport. As you might expect, industrialized countries produce the most CO_2. Unfortunately, though, once it is in the atmosphere CO_2 spreads out over the whole world, so it causes a problem everywhere, not just over the people who produced it.

> ***'It is now clear that man-made greenhouse gases are causing climate change. The rate of change began as significant, has become alarming and is simply unsustainable in the long term.'***
> UK Meteorology Office

Not just warmer days

The rise in temperature has knock-on effects that are more damaging than just warmer days. As ice melts in the north and south polar regions and high in the mountains, more water runs into the oceans, causing the sea level to rise. Ice sheets reflect more of the Sun's heat than does liquid water in the sea, so removing ice leads to temperatures rising further. Warmer water can absorb less CO_2 than colder water, so as sea temperatures rise the oceans hold less dissolved CO_2, leaving more in the atmosphere which leads to further temperature rise. Since 1992, an average of 65,000,000 tonnes of ice have been lost from Antarctica every year.

Many climate scientists believe there could be a tipping point beyond which the melting of polar ice can't be stopped. A vicious cycle of rising sea levels leading to higher temperatures could

continue until all the ice has melted, giving a final sea level 65m (213ft) higher than it is today. The last time the Earth was ice-free was 30 million years ago, long before humans evolved.

As sea levels rise, coastal areas will flood. A warmer climate will lead to more droughts, changing patterns of crop growth and water shortages in areas that depend on melting glaciers for their water supply. Whether the effects will be catastrophic for humans depends on how much temperatures and sea levels rise.

Climate change denial

Some people reject the theory and its supporting evidence. It is certainly inconvenient, as it puts the onus on us to change our behaviour and stop the warming if we want to avoid its consequences. There are two types of dissenter: those who deny there is a problem with warming and those who deny that human activity is causing climate change (but don't deny it is happening or causes harm).

Those who deny there is a problem with warming generally argue that it will be good for plant life, which could remove CO_2 from the atmosphere, and that life on Earth has existed at higher temperatures before. The latter is true, but life at significantly higher temperatures didn't include humans. Rising sea levels will lead to flooding of much of the land we currently occupy and rely on for farming. The crops we grow for food will not tolerate large changes in climate and some farmland will be flooded while other areas will be parched by drought. Last time it was warmer, some low-lying areas that are now densely inhabited, such as London and Hong Kong, were flooded.

Those who deny that human activity is causing climate change point to other sources of warming, such as volcanic activity, the Sun's activity and changes in the Earth's orbit. Currently, the Sun is cooling slightly and the Earth's orbital position would tend to make the world colder, so these factors are unlikely to be warming the Earth. Also, the amount of CO_2 added to the atmosphere by volcanoes does not change much over time and is only about one hundredth of the amount added each year by human activities.

Some sources of CO_2 are outside human control. We don't cause volcanic eruptions, alter the orbit of the Earth, move the continental landmasses around or change the activity of the Sun. But we do choose how to use the land, control how many plants grow, farm extra animals and burn a lot of fossil and other organic fuels. The trend of human activities is to clear land of natural vegetation and add CO_2 to the atmosphere. When dense forest is replaced with grassland for cattle, the same area of land removes far less CO_2.

Up and up

The level of CO_2 in the atmosphere has been measured continuously since 1960 from a research station at Mauna Loa in Hawaii. There has been a steady rise, though each year the concentration swings between high and low values corresponding to winter and summer in the northern hemisphere (where most people live). The higher value in winter is related to burning more fossil fuels to provide heating and lighting during the colder and darker time.

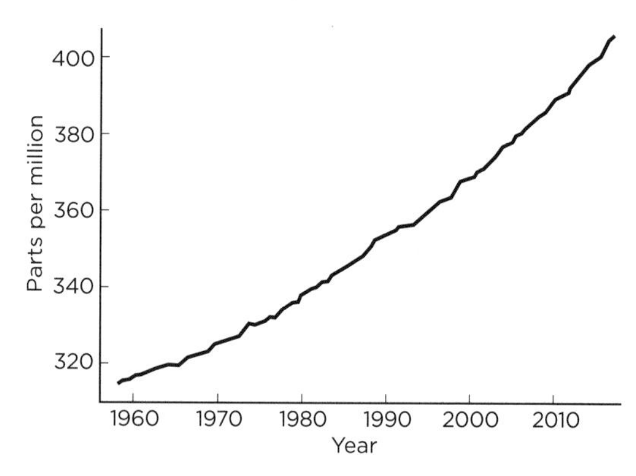

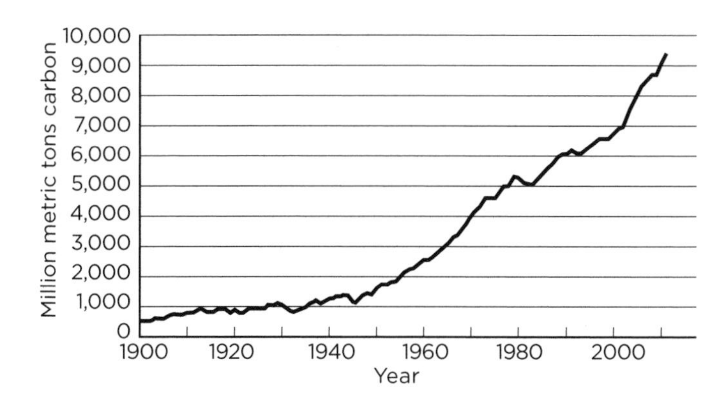

The concentration of CO_2 has risen at an increasing rate since 1900, going up less than 1,000 tonnes in the 50 years from 1900 to 1950, but 5,000 tonnes in the following 50 years (1950–2000) and 3,000 tonnes in the next ten years (2000–2010).

Last time round

The last period of rapid global warming happened 55 million years ago. The average temperature rose by 5° to 8°C over a period of 200,000 years when massive volcanic eruptions poured far more CO_2 into the atmosphere than usual. If climate change theory is correct, we could see a rise of 5°C in less than 100 years. It would be very hard for living organisms – and human societies – to adapt to such a rapid change.

Chapter 6

Germ Theory

There have been many ideas about what makes us ill, but germs were not on the list until after the invention of the microscope.

THE IDEA

Many illnesses are caused by the action of micro-organisms. They invade a living host and reproduce, causing disease.

Demons and deities

Illness can seemingly appear from nowhere – one day you are healthy, the next you are sick. In an age when belief in supernatural entities was the norm, it would have been easy to think a quixotic god or demon had picked on you to be sick, either randomly or in retribution for something you might have done.

If you lived in a culture that blamed illness on supernatural beings, it would seem reasonable to seek recovery by supernatural means – through prayers, incantations, offerings, sacrifices, exorcisms, rituals or other methods of appeasing gods or driving away demons and spirits. Inevitably, some of the people who used those methods would recover. Then it would look as though the method had worked. No one would stop to think that they might have got better anyway.

It wasn't just individuals who could be

The ibis is a bird that eats water snails. The presence of ibises at a fishpond in Ancient Egypt prevented parasitic diseases that could be passed on through water snails. Although the Egyptians did not know this mechanism, they revered the ibis.

punished with a nasty illness; an angry deity could blast a whole community – even the whole world – with a plague. The Black Death was an epidemic, probably of bubonic plague, in the middle of the 14th century which killed up to half of the population of Europe and Asia. In Christian Europe, people were quick to claim that God had sent it as a punishment for the corruption rife in the world. To try to appease God and drive away the pestilence, they resorted to prayers and penance.

Is it catching?

Even with no knowledge of medicine, it's easy to divide illness into two categories: diseases that you catch from someone else and those that just arise in your own body. If someone in your household catches a cold, you stand quite a good chance of catching one too. But if someone in your household develops heart disease or lung cancer, you are not going to catch it from them any more than you would catch a broken leg.

As people began to look for physical causes for diseases, the distinction between those that can be caught and those that cannot was significant. This began to happen first in Ancient Greece, around the 5th century BC.

The unbalanced body

Diseases which arise spontaneously within the individual were often explained in terms of an imbalance of the four humours or body fluids (see page 160). This idea was promoted by the Roman physician Galen in the 2nd century AD and remained current until the 19th century. Diseases that spread through a population are a bit harder to explain in these terms, though. Some diseases were associated with environmental

The three great ancient teachers of medicine: Galen (Roman), Avicenna (Persian), and Hippocrates (Greek).

conditions that were thought to affect the balance of the humours, so cold, wet weather could lead to rheumy conditions in which people produced too much phlegm, for example. This could explain why lots of people get colds and flu in the winter. But not all diseases could be explained in terms of environmental conditions upsetting the balance of the humours. Other models were needed.

'Bad air'

A common explanation for diseases we now consider infectious was 'bad air' or 'miasmas'. The 'bad air' was thought to be produced by decaying organic matter, so anywhere that decaying material was found was considered unhealthy. Swamps, battlefields and forests were obvious contenders and – reinforcing the idea – were also places where people fell ill. Malaria used to be common in the swampy land around the Veneto in northern Italy. It is caused by a parasite carried between people by mosquitoes. Rather than making a connection with the mosquitoes that bred in the swamps, people in the region blamed the air around the marshy land. Indeed, the name 'malaria' means 'bad air'.

It was easy to make connections between poor environmental conditions and illness. As slums spread through European and North American cities in the 19th century, health worsened in the crowded, damp, dirty conditions. It made sense in the light of the 'miasma'

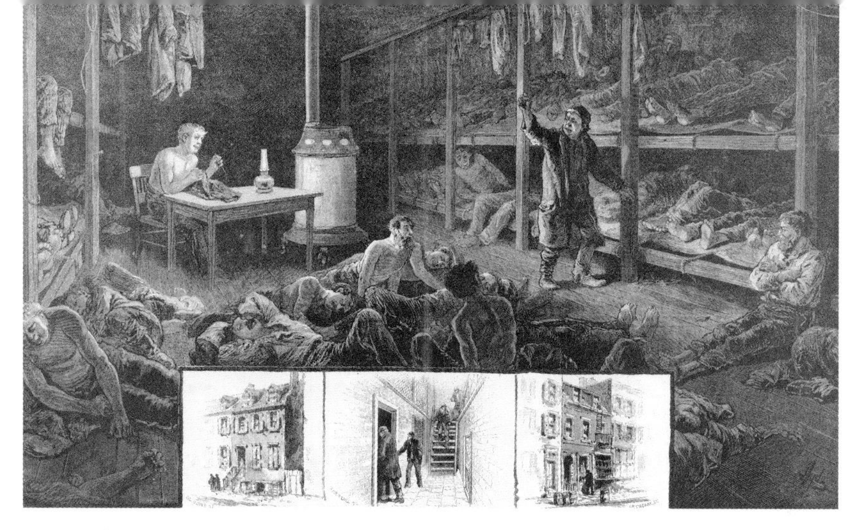

Many of the urban poor lived in deplorable conditions in the 19th century, and disease was rife.

model of disease: the air was rank and wet, or sometimes too hot, there was rotting rubbish and excrement on land and in the rivers and people naturally fell ill. When London was plagued by smog and lung diseases spread rapidly through the city in the 19th century, 'miasmas' were blamed.

Looking ahead

There might have been something like a germ theory circulating in the 1st century AD. In a book on agriculture, Roman writer Marcus Terentius Varro included the advice that people should not build

Fracastoro suggested tiny mobile particles might cause disease.

houses near swampy land because there are tiny creatures too small to see which 'enter the body from the air through the mouth and nose and cause serious diseases'. If anyone else shared this idea, they left no record of it.

Fourteen hundred years later, in 1546, Italian physician Girolamo Fracastoro suggested that epidemic diseases could be caused by tiny 'spores' which can be moved around on fabric or even act at a distance to spread disease. It's not clear whether he considered them to be chemicals or living entities.

The start of something small

In the 1590s, Hans Jansen and his son Zacharias, two spectacle-makers from Holland, experimented with arranging magnifying lenses in a tube. They found that something placed at the end of the tube looked much larger through the series of lenses than through just one lens. They had invented the compound microscope.

At first, microscopes offered quite low powers of magnification. It was possible to see in detail small animals such as fleas and even the protozoa found in pond water, but it was not possible to see anything as small as bacteria.

Still, once really small things had been seen clearly, including some too small to see at all with the naked eye, conjecture was possible. In 1650, August Hauptmann suggested that tiny creatures, 'animalcules', could perhaps cause disease. He thought that either the creatures – a type of worm – or their eggs might spread contagion.

A few years later, Athanasius Kircher made the same proposal, but more fully articulated. He suggested that the plague could be caused by 'hidden seeds of a deadly nature' that infected plague victims. When the infected person breathed out, they spread some of these poisonous seeds into the air

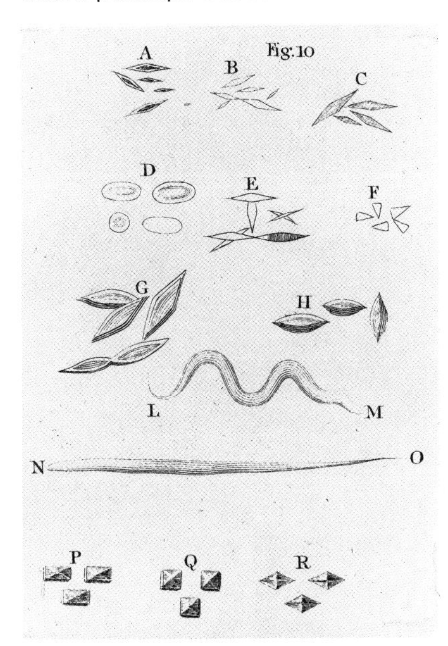

Microscopic animalcules seen in white wine vinegar by Anthony van Leeuwenhoek, 1798.

and warmth gave them life, so that they might infect someone else. He warned that the particles could be carried between people in many ways, including on the breath, deposited on objects such as clothing and surgical instruments, by physical contact or through contaminated food or drink. He claimed to have seen microscopic 'worms' in the blood of plague victims.

The first person to see bacteria, in 1676, was the Dutch microbiologist Antony van Leeuwenhoek, who made his own microscopes. His instruments were more powerful than any made previously and enabled him to see for the first time sperm, blood cells, bacteria and much else. Seeing them, though, was not the same as knowing what they did. One interpretation was that the disease caused the micro-organisms to appear in blood or other body fluids, rather than that their presence caused the disease.

Germ theory tried out

Frequent epidemics of cholera swept through the city of London in the 1840s and the prevalent belief was that miasmas from the River Thames were to blame. The disease is characterized by watery diarrhoea and caused by a bacterium. The physician John Snow, sceptical about miasmas, suspected instead that there was some physical agent of disease behind the epidemic. Although germ theory was still in its infancy, in 1849 he proposed a mechanism

An itinerant microscopist showing animalcules to children in Friesland.

by which infection could be transferred between people, carried in faeces and consumed so that it infected another person's gut: 'The excretions of the sick at once suggest themselves as containing some material which being accidentally swallowed might attach itself to the mucous membrane of the small intestines, and there multiply itself by appropriation of surrounding matter, in virtue of molecular changes going on within it, or capable of going on, as soon as it is placed in congenial circumstances.'

By mapping all incidences of the disease in an outbreak in 1854 he managed to trace its source to a water pump that was drawing infected water. His recommendation that all water be boiled before

PROTECTION AGAINST THE UNKNOWN

Just as Snow found a method of protecting people against the cholera bacterium before it was discovered, the English surgeon Joseph Lister found a way of protecting surgery patients against the bacteria that cause wound infections before knowing what they were. He first tried carbolic acid on a wound dressing while treating a young boy with a compound fracture of the leg. Compound fractures (in which the bone sticks out through the skin) almost always became infected and most victims died. Lister found his patient recovered with no infection. He then began washing his surgical instruments in a 5 per cent solution of carbolic acid and sprayed the solution in the operating theatre during surgery, vastly reducing the rate of infections in his hospital.

Lister's discovery was a breakthrough. Reducing the very high risk of infection enabled far more complex surgery than had previously been possible. Soon after, when Pasteur demonstrated the action of micro-organisms, it became clear how antisepsis works to prevent infection by killing bacteria.

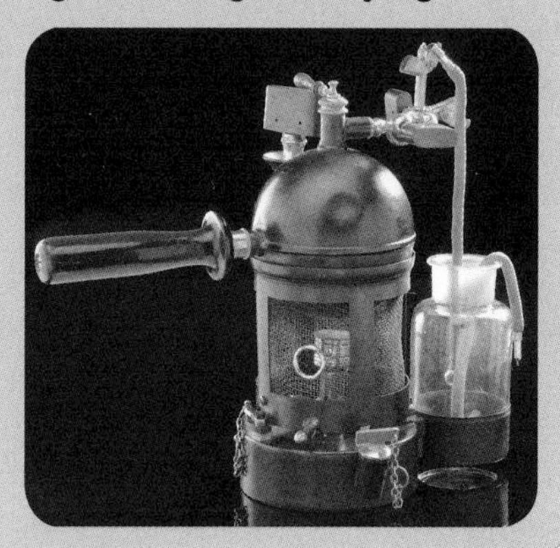

A spray for carbolic acid, used to make surgery safer.

drinking was the first practical application of germ theory to combat disease. Although his explanation was not immediately accepted, in the second half of the 19th century many major cities invested in keeping water supply and sewage removal systems clearly separate. Consequently, cholera epidemics were virtually eradicated in the developed cities.

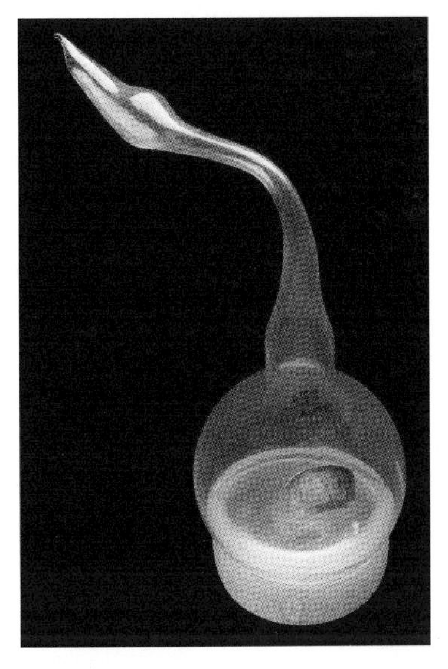

In a sealed flask, no bacteria grow in boiled broth.

Bacteria and broth

In 1856, French chemist and microbiologist Louis Pasteur was asked by a wine merchant in Lille for help with a problem he had – making and keeping alcoholic drinks without them spoiling. Pasteur showed that the fermentation process is carried out by a micro-organism (yeast) which must be present in the brew for sugar to be converted into alcohol. He also found that if the liquid was contaminated (with bacteria) it could spoil. The bacteria were producing lactic acid. Pasteur showed, through his broth experiment, that bacteria are also responsible for spoiling milk, broth and other foods (see page 33).

From drinks to disease

Having demonstrated the action of micro-organisms in spoiling food, it was logical to speculate that they might also cause disease – especially as disease was often associated with decaying or rotted matter, including bad food.

Pasteur worked on cholera in chickens, discovering that he could infect healthy chickens with matter taken from sick chickens. He went on to develop the first laboratory vaccine, finding (accidentally) that a weakened form of the bacterium could make chickens ill, but that they then recovered and were immune to further doses of full-strength cholera.

Mapping microbes to diseases

After deciding that micro-organisms can cause disease, the next step was to find the micro-organism that causes a particular disease. German microbiologist Robert Koch set about doing this in the 1870s and 1880s. Working in a laboratory he had built in the countryside outside Berlin, he turned first to anthrax, a disease that affected farm animals. The previous century, anthrax had killed half the sheep in Europe – it was a serious threat.

> ***'As soon as the right method was found, discoveries came as easily as ripe apples from a tree.'***
> Robert Koch

Koch wanted to demonstrate that the bacterium *Bacillus anthracis* causes anthrax. He took some of the bacteria from a sheep that had died from the disease and injected it into a mouse. The mouse developed anthrax. He took bacteria from the mouse and infected another mouse. He carried this out again and again until he had infected 20 mice in sequence and then, in 1876, he announced that he had identified the bacterium which causes anthrax. He went on to identify the bacteria which cause tuberculosis and cholera.

In 1890, Koch set out four 'postulates' or criteria that must be met before a particular micro-organism can be considered to cause a particular disease.

- The micro-organism must be present in all cases of the disease.
- It must be possible to take the micro-organism from a diseased host and grow it in a culture medium.
- The cultured micro-organism must produce

Louis Pasteur inoculating a sheep against anthrax.

the same disease when used to infect a previously healthy animal.

- The same micro-organism must be found in the new host.

Smaller than very small things

While Pasteur had been able to see bacteria with his microscope, there was another type of pathogen (micro-organism which causes disease) lurking that was too small to see. The Russian botanist Dmitri Ivanovsky, working with the tobacco mosaic disease, discovered in 1892 that he could use sap from diseased plants to infect healthy plants. He found he could do this even after filtering the liquid to remove all solid particles. His porcelain filter was fine enough to remove bacteria, so he realized there must be something else that was carrying the disease. Among his suggestions was that bacteria had produced a toxin that remained in the sap. The pathogen could not be cultured, however, so Koch's postulates could not be applied.

Six years later, in 1898, the Dutch microbiologist Martinus Beijerinck independently made the same discovery and claimed the new infectious agent was a 'contagious living fluid'. Ivanovsky had seen crystal-like growths in cells of the diseased plants, but Beijerinck thought the pathogen was a liquid. A debate about whether viruses were liquid or particles continued for 25 years. In just a few years, though, knowledge of viruses increased rapidly, with several others having been found – first the one which causes foot and mouth disease (in 1898) and later the first human virus, yellow fever (in 1901).

The tobacco mosaic virus was first crystallized by the American biochemist Wendell Stanley in 1935 and viewed with an electron microscope in Germany the same year. The discovery that the virus was essentially a chemical – needle-shaped crystals with the properties of a protein – was astonishing: how could something that can reproduce and cause infection be an inert chemical? It was soon found to be a combination of protein and RNA (see page 262). This raised the question of whether viruses can be considered alive when they are not cellular and can't reproduce or 'live' outside a host. It prompted discussion of how we define life. Viruses still seem to straddle the living/not-living border, sharing some features with living organisms (such as being able to reproduce themselves) and lacking others (such as growing and making their own energy).

Beijerinck was among the first people to recognize and work with viruses.

Chapter 7

Geocentric & Heliocentric Theories

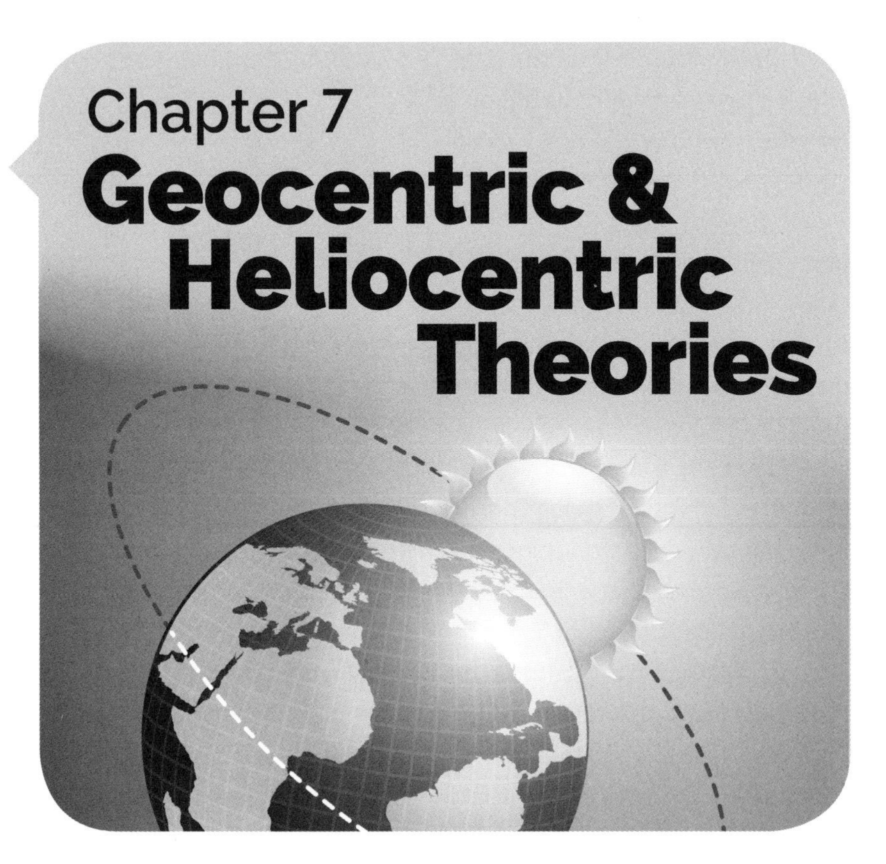

Every day you can see the Sun rise in the east and set in the west. It looks as though the Sun is going round the Earth. How do we know it's not?

Looking upwards

It might seem obvious to us that the Earth goes round the Sun, but that's only because we have grown up with the idea. In fact, there is no easy way of telling which body is in orbit and which is the focus of the orbit – it would look the same from Earth either way. Two theories have tried to explain the structure of the solar system and the motion of bodies within and outside it: the Ptolemaic and the Copernican models.

THE IDEA

There have been two important competing theories about the movement of the Earth, planets and Sun.

- **Geocentric theory says that the Earth is fixed at the centre of the system and the Sun, Moon and other planets orbit the Earth.**
- **Heliocentric theory says that the Earth and other planets all orbit the Sun.**

As good a guess as any

The Ancient Greek philosopher Anaximander (*c*.610–546BC) described a cylindrical Earth floating unsupported in space and circled by the heavenly bodies that we can see – the Moon, Sun, planets and stars.

The opposite view – that the Sun is central and the Earth moves – was first proposed by Aristarchus of Samos (*c*.310–*c*.230BC). He had the Earth rotating daily on its own axis and orbiting the Sun over the course of a year. He believed that the other planets and the stars also orbit the Sun, stacked in concentric circles or spheres around it. He also proposed that the stars are other suns, and the Sun is a star, but most stars are too far away for us to feel their heat. Their great distance is also why they don't seem to move relative to one another.

Like Anaximander before him, Aristarchus looked at what he saw in the night sky and tried to work out a model that would explain its appearance. Neither had a telescope nor any way of telling which was the right model, but Aristarchus was the one who picked the right option.

Not everyone was convinced by Aristarchus's ideas, though. In fact, most people weren't. The dominant model put the Earth at the centre. It was promoted by the great Greek philosopher Aristotle in the 4th century BC, and developed in the 2nd century AD by the Egyptian-Greek Ptolemy. The Ptolemaic theory of the universe dominated Western astronomy for 1,500 years.

The problem with planets

If the planets just moved in circles round the Earth, a geocentric model of the solar system would be very persuasive. But they don't seem to. Observed from Earth, the planets follow a strange, wandering path. Indeed, the name 'planet' comes from the Greek for ‘wanderer’. The planets seem to follow a series of loops, and not even regularly spaced loops. Each planet appears to move forwards, then stops, then goes backwards for a bit, then forwards again.

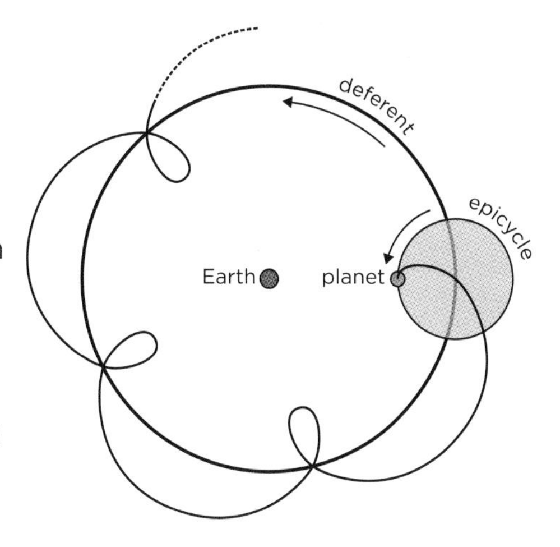

In the 3rd century BC, Greek astronomer Apollonius of Perga suggested that each planet’s path is defined by two circles rather than just one. The planet travels round in a small circular orbit, called an epicycle. Then this epicycle itself, as a whole, orbits the Earth. The larger circle is called the deferent.

Even this did not quite explain the movement, though, as the loops are not evenly spaced or sized. It helped to offset the planet’s orbit so that the Earth was not central; but this meant the planet’s movement was neither perfectly circular nor at a uniform speed, which Greek philosophy demanded.

Ptolemaic theory

Ptolemy tackled the problem of the apparently rather chaotic motion of the planets in a clever way. More importantly, he came up with a mathematical explanation of the movement of the Moon, Sun and planets round the Earth. Once he had a mathematical model, it could properly be called a theory as it would be possible to predict the future position of a planet.

The centre of the deferent is called the eccentric (there is nothing at that point since Earth is offset from the centre). Ptolemy added another point to the model, which he called the equant. This was Earth's mirror image, the same distance from the eccentric as Earth but on the opposite side. Ptolemy calculated that a planet's speed was uniform if observed from the equant – but not if observed from anywhere else, including from Earth or the eccentric. The speed he was talking about was angular speed, meaning that the planet would move through the same angle of arc across the sky in the same interval of time.

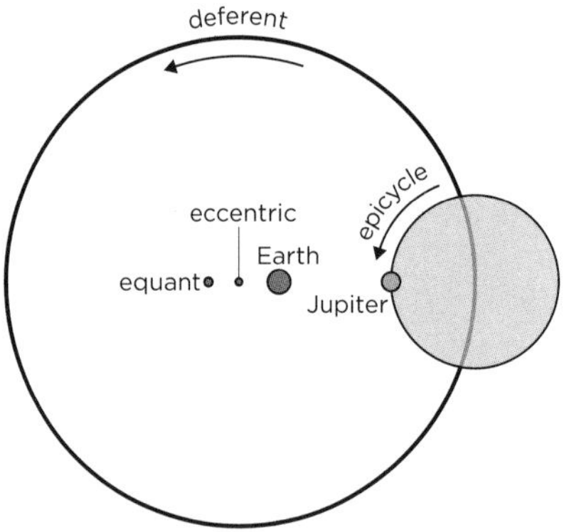

Circles within circles

In Ptolemy's theory, each of the planets, the Moon and the Sun occupied a sphere or 'orb'. The stars (called the 'fixed stars' to distinguish them from the wandering planets) all occupied a single orb, the furthest from Earth. Ptolemy put the Moon closest to the Earth – that much was clear as it can pass in front of any of the others, obscuring planets and stars or causing an eclipse of the Sun. The fixed stars had to be most distant, as the others move relative to them. The movement of the planets and the Sun was measured against the band of stars which make up the constellations of the zodiac, a circle of background stars, different ones being visible from Earth at different times.

Obviously, we can only see the stars at night-time, so when we look at the night sky we see those that are in the opposite direction to the Sun. For Ptolemy, though, which stars were visible depended on the rotation of the orb of fixed stars.

All in order

The time a planet takes to make a complete circuit of the zodiac, as viewed from Earth, is known as its sidereal period. The outer planets (those further from the Sun than Earth) have progressively longer sidereal periods. This gives

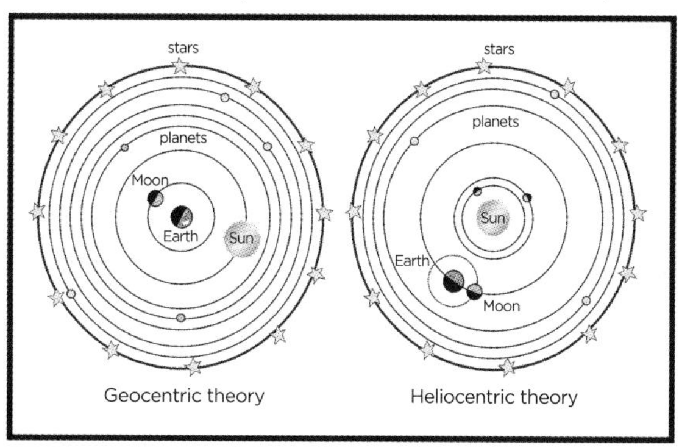

a sequence for those planets in the Ptolemaic model: in order of distance from Earth they must be Mars, Jupiter and Saturn. As Mercury, Venus and the Sun all have a sidereal period of one year, there's no obvious way of choosing the order they should go in. Ptolemy followed tradition in putting them in the order Mercury, Venus, Sun (with Mercury closest to Earth).

With the planetary motions explained adequately, it looked as though Ptolemy's theory was correct – there was no immediate need to challenge it.

Drifting . . .

However, using Ptolemy's calculations, the planets seemed to drift off course, moving further from where they should be as time passed. It was a slow drift, over centuries, so it took a while for people to notice. Several Arab astronomers refined Ptolemy's work and began with new positions, but it was not entirely satisfactory. No one was willing to overturn the model completely for a very long time – perhaps because they had nothing to replace it with.

Reinventing the universe

Despite its problems, Ptolemy's theory prevailed until the 16th century. Then, in 1543, the year of Ptolemy's death, Polish astronomer Nicolaus Copernicus published a radical new theory. It ousted

TILTING EARTH

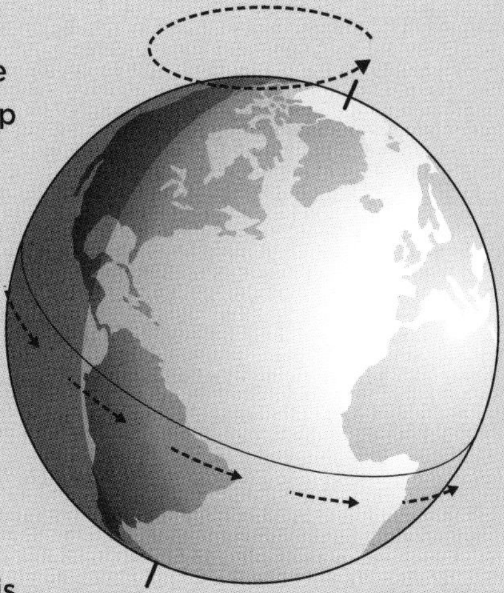

The Earth's axis is at a tilt, so the North Pole is not really at the top (though 'top' doesn't mean much in space). In effect, this means that the axis (the line from the North to the South Pole about which the Earth rotates) is not at 90 degrees to the plane of the Earth's orbit around the Sun. It is the tilt that produces the seasons as the Earth orbits the Sun.

The axis is not only tilted but is very slowly moving. If we extended the axis from the North Pole out into space, it would describe a circle, completed over a period of about 26,000 years. This is called axial precession, or the precession of the equinoxes.

The apparent positions of the stars gradually change over this period as we are looking at them from a slightly different angle. By the end of 26,000 years, we are back where we started and they look the same again. As the Ptolemaic system measures movement against the fixed stars, it is something of a problem that the stars are not actually fixed with respect to Earth. If we waited another 24,000 years, Ptolemy's scheme would look nice and accurate again.

Earth from its place at the centre of the universe and demoted it to the position of a planet orbiting the Sun along with the others. *De revolutionibus orbium coelestium* (*On the Revolutions of the Celestial Spheres*) was one of the most important scientific books ever published, but like so many world-changing new theories, it met with resistance.

Copernicus's theory puts the Sun at the centre of the solar system and, probably, of the universe. It is based on seven assumptions:

1 There is no single centre to all the celestial circles or spheres.

2 The centre of the Earth is not the centre of the universe; it is just the centre of its own gravity and of the Moon's orbit (the lunar sphere).

3 The Sun is at or near the centre of the universe; the planets, including Earth, orbit round it.

4 The distance from Earth to the fixed stars is so much greater than the distance from Earth to the Sun that the Earth–Sun distance is imperceptible in comparison.

5 Whatever motion appears in the firmament (the region of the fixed stars) arises not from the firmament moving, but from the Earth moving. The Earth rotates on its fixed poles (its axis) every day, while the highest parts of the heavens remain unchanged and unmoving.

6 The apparent motion of the Sun is the result of the Earth moving, not of the Sun moving. The Earth rotates on its axis and revolves round the Sun in the same way as any other planet. Earth therefore has more than one type of motion.

7 The apparent retrograde (backward looping) movement of the planets is the result of Earth's movement. The motion of the Earth explains many apparent inequalities, foibles or imperfections in the movement of heavenly bodies.

Copernicus lectures to other astronomers.

Fighting back

Relinquishing Earth's place at the centre of the universe was difficult at a time when most Europeans believed that humans were God's supreme creation. A geocentric theory of the universe seemed to be in tune with God's plan; it was certainly the model preferred by the Church. It didn't seem reasonable that God would have put his finest creation on one of several planets orbiting the Sun rather than in a special position. Opposition to Copernicus's theory began immediately.

An agent of the Lutheran Church, the minister Andreas Osiander, wrote a preface which was attached to *De revolutionibus* at its first printing. The preface appeared without attribution, so it would seem to be Copernicus's own work. Copernicus at the time was too ill to

A cosmological map showing Copernicus's astronomical vision.

THE CENTRE OF THE UNIVERSE?

Some people were not ready to give up Earth's position as central to the universe, but were persuaded that having planets orbit the Sun was a good idea. One of these was Tycho Brahe, a flamboyant and eccentric astronomer who had an observatory on a tiny island in Copenhagen, Denmark. Forty-five years after the publication of Copernicus's book, he suggested that the Moon and Sun orbit the Earth, but all the other planets revolve around the Sun.

do anything about it, even if he knew. The preface tried to reduce the impact of the book by suggesting the theory was just that – entirely theoretical, a model which would help to calculate and describe the positions of the heavenly bodies, rather than reflecting beliefs about the genuine nature of the universe.

This might seem a strange thing to do, but in 1277 the Church had set out a list of 219 ideas which were condemned as heretical. Anyone who held or promoted them could be excommunicated. Two 14th-century philosophers, Jean Buridan and Nicole Oresme, came up with a workaround which allowed scientists to propose ideas that conflicted with the Church's teaching – but only as long as they didn't suggest they were actually true. Scientists could propose a hypothesis in keeping with observed phenomena as long as it was treated only as an idea and was not trying to constrain God to work in certain ways. Osiander's preface to Copernicus's treatise presented it as a model for thought, rather than an attempt to explain how the universe really is structured.

> ***'On no point does [Copernicus's theory] offend the principle of mathematics. Yet it ascribes to the Earth, that hulking, lazy body, unfit for motion, a motion as quick as that of the aethereal torches, and a triple motion at that.'***
> Tycho Brahe

Increasingly unpopular

Although a few astronomers found Copernicus's theory compelling, the Church was not convinced. It had tolerated his book, with its preface, but as other astronomers began to teach the theory as though it really did explain the universe, the Church became increasingly uneasy. In 1616, *De revolutionibus* was banned and the Italian scientist Galileo Galilei was instructed not to 'hold, defend, or teach' heliocentrism.

Many astronomers rejected the theory. They were unwilling to allow the Earth to move, either on its axis or around the Sun. Furthermore, Copernicus's model did not produce more accurate predictions for the positions of the heavenly bodies than Ptolemy's model. The one thing in its favour was that it returned the planets to proper circular motion with no need for epicycles. (Epicycles were still used to calculate their position, though, as this was the best way of modelling mathematically the observed movements of the other planets.)

An observatory in western Europe in the era of Copernicus.

There was really no way to decide between the two systems based on empirical evidence: it was simply a choice of the established system that seemed to follow common sense (why assume the Earth moves when it doesn't seem to?) or one with simpler geometry. It took until the 17th century before other astronomers were convinced by Copernicus.

Taking Copernicus further

One astronomer who was persuaded by the Copernican model was Johannes Kepler (1571–1630). He had worked with Tycho Brahe and taken over Brahe's job and data on his death in 1601. Up until this point, everyone assumed the orbit of the planets, whether round

Kepler completed the Copernican model; his work still stands.

the Sun or the Earth, was circular. Circular movement was, after all, what the Greek philosopher Aristotle had described as the perfect movement and there was no reason to doubt it.

But Kepler, with the advantage of Brahe's daily observations of the position of Mars, discovered that if he plotted the course of this planet it was an ellipse with the Sun at one focus. This was the breakthrough that was needed. When Copernicus's theory was refashioned using elliptical orbits, everything fell into place and the discrepancies between calculated and observed positions all but disappeared.

Kepler published his findings in 1609. The following year Galileo published his own discovery, made with the newly invented telescope, that four moons orbit the planet Jupiter. This really threw the cat among the pigeons. The Ptolemaic model had no space for any bodies orbiting anything other than Earth. Even if Jupiter were still allowed to orbit Earth with its moons, something fundamental to the theory had to give way. Furthermore, Aristotle had specified that there is no empty space in or between the orbs of the heavenly bodies (he didn't believe an empty void could ever exist), so nothing could be moving round within the orb of Jupiter.

Held in place

In providing more accurate predictions, Kepler's revisions to Copernicus's theory persuaded some astronomers that it could be correct. But Kepler wasn't able to explain how or why the planets might stay in orbit round the Sun. Aristotle had proposed orbs made of a special substance, celestial aether; he also said they had a natural circular motion and once set turning would turn forever. This was plausible if the planets really were describing circles, but less compelling if they moved in elliptical orbits. The answer was provided in 1687 by Isaac Newton. His theory of gravity (see page 114) explained how a moon could be held in orbit round a planet and a planet round a sun or star.

Copernicus proved correct

And the final proof? Not until the space age could we finally provide physical, rather than mathematical, proof in support of Copernicus's theory. With cameras and astronauts in space, it was finally possible to watch the Earth and the planets move on their paths around the Sun.

Chapter 8

Theory of Gravitation

There is more to gravity than falling, though that's its most obvious manifestation and the one that first attracted attention.

Where is down?

> **THE IDEA**
>
> **Bodies with mass exert a force – gravity – which attracts all other bodies with mass. The force between two objects is proportional to their masses multiplied together and inversely proportional to the distance separating them.**

You don't have to be very observant to notice that things fall downwards: towards the ground, from a higher elevation to a lower elevation, or from the surface into a hole. As soon as people thought about the Earth as a globe (and that was a long time ago), it must have been clear that 'down' really means 'towards the centre of the Earth'. If we could make a hole all the way through the Earth and out the other side, we'd quickly find that things dropped into the hole would come to a stop at the centre of the Earth – they wouldn't go all the way through.

The distinction between 'down' and 'towards the centre of the Earth' becomes more important when we start to think about gravity in space rather than just on Earth.

Galileo's falling objects

There's a popular story that Galileo once dropped a heavy ball and a light ball from the Leaning Tower of Pisa to show that they both hit

The Leaning Tower of Pisa, site of Galileo's alleged experiment.

the ground at the same time. One of Galileo's pupils reported that the experiment took place in 1589, but Galileo made no reference to it so most historians consider it a myth. If it did take place, it would have been to demonstrate a point about falling objects.

Aristotle said that heavy objects fall more quickly than light objects dropped from the same height. Intuitively, this seems sensible. We would expect gravity to exert a greater force on a heavier object as it has more to work with. Indeed, everyday experience often seems to bear this out: if you drop a cricket ball and a feather from a window, the ball will hit the ground first. This isn't because it's heavier. It's because there is more drag,

> ***'Let us take . . . two balls of lead, the one ten times bigger and heavier than the other, and let them drop together from 30 feet high, and it will show, that the lightest ball is not ten times longer under way than the heaviest, but they fall together at the same time on the ground. . . . This proves that Aristotle is wrong.'***
> Simon Stevin, Flemish mathematician, 1586

or air resistance, acting on the feather and the effect of air currents is greater on the feather than on the ball (because it's lighter and its shape is less aerodynamic). You can test this easily by dropping two identical sheets of paper, one flat and the other screwed tightly into a ball. Their weight is the same, but the ball is less hampered by the air.

Although Galileo may not have tried this experiment, several other people did, with reports as early as 1544 of tests demonstrating that the heavier object doesn't fall more quickly. In 1576, Giuseppe Moletti (Galileo's predecessor as professor of maths at the University of Padua) wrote that when both heavy and light bodies of the same material, and bodies of equal weight but different materials, are dropped together they hit the ground at the same time.

The law of falling bodies

After about 20 years' work with falling bodies, Galileo formulated a law of motion which said that in a vacuum all bodies, of any weight

GALILEO'S PARADOX

In his unfinished book *De motu* (*On Motion*), Galileo described a thought experiment in which a heavy ball and a light ball are connected by a string and dropped from a tower. If the heavy ball is inclined to fall more quickly the string will become taut. Suppose you start with the theory that the rate of fall is directly related to the weight of the object and something that weighs more will fall more quickly. Tying the balls together should cause the slowly falling light ball to impede the fall of the quickly falling heavy ball. But the pair of balls tied together will be heavier than the heavy ball itself, so surely they should fall faster? The contradiction suggests that the assumption is wrong.

or shape, accelerate in exactly the same way, and that the distance fallen is proportional to the square of the time it takes them to fall. His law would soon find much wider application.

Newton, the apple and the Moon

It's all very well to say that gravity causes things to fall towards the centre of the Earth, but alone that's an observation rather than a theory. The first person to suggest the reason in a way that has universal application was the English scientist Isaac Newton (1642–1727). According to his assistant, John Conduitt, the idea came

to Newton as he sat in the garden of his mother's house in Lincolnshire in 1666, having been forced to leave Cambridge because of the plague:

'It came into his thought that the power of gravity (which brought an apple from a tree to the ground) was not limited to a certain distance from earth, but that this power must extend much further than was usually thought. Why not as high as the Moon thought he to himself & that if so, that must influence her motion & perhaps retain her in her orbit, whereupon he fell a-calculating what would be the effect of that superposition.'

It was the realization that gravity extends far beyond the surface of Earth, and is in fact a universal force that operates everywhere, that was groundbreaking.

The idea that Earth's gravity might extend as far as the Moon could be tested only mathematically at the time, as it was not possible to send an object even to the edge of the atmosphere.

Explaining gravity

Newton both proposed a theory and derived a law which allowed him to quantify gravity and predict the force acting between any two bodies (objects). His theory was that gravity is an attractive force which operates between any two bodies with mass: not just between the Earth and something falling towards it, but also between Earth and the Moon and the Sun and Earth.

It was clear to Newton that the strength of the attraction must be related to both the mass of the bodies and the distance between them. This means not only does Earth exert a pull on you, which keeps you on its surface, but you exert a pull on the Earth. The same is true of every possible pair of objects, though usually we only notice the force that the more massive (heavier) object exerts. We don't notice the gravitational pull between, say, an apple and a table because the gravitational pull of the Earth over both of them is so much greater that it's all we see.

Newton came up with a mathematical rule for calculating the force

$$\text{Gravitational force} = \frac{\text{Gravitational constant} \times (\text{mass of body 1}) \times (\text{mass of body 2})}{(\text{distance between bodies}) \times (\text{distance between bodies})}$$

of gravity between two objects, expressed in the formula:
This is written as:

Gravitational force = (G x m1 x m2) / (d^2)

The gravitational constant (G) is a fixed number with the value of 6.673×10^{-11} $N{\cdot}m^2/kg^2$. A Newton (N) is the force needed to accelerate a mass of 1kg (2.2lb) at 1m (3.3ft) per $second^2$ in Earth's gravitational field.

MEASURING ACCELERATION

Speed (or velocity) is a measure of distance/time - how far something travels in a set interval of time. We measure vehicle speeds in kph (kilometres per hour) or mph (miles per hour). In science, velocity is usually measured in metres per second (m/s) or centimetres per second (cm/s).

Acceleration is a measure of the rate of increase in speed, in other words, how quickly something speeds up. If an object's acceleration is given as 1cm per second per second, or $1cm/s^2$, this means the velocity increases at the rate of 1cm per second every second. Therefore after 1 second, the velocity is 1cm/s, after 2 seconds it's 2cm/s, after 10 seconds it's 10cm/s, and so on.

Gravity and acceleration

If you drop an object, it accelerates downwards. That means it travels more and more quickly until it hits the ground. (If it falls from high enough, it stops accelerating at a certain point – see box opposite). On Earth, the acceleration due to gravity (in a vacuum) is always $9.8m/s^2$. This means that after 1 second, the object is travelling at 9.8m/s, after 2 seconds it's travelling at 2 x 9.8 = 19.6m/s, and so on.

We can think of gravity as the pull that wants to accelerate us towards the centre of the Earth. As we are on the solid surface of the Earth, it just stops us flying off rather than pulling us underground.

The same force is pulling the Moon towards Earth, but the Moon is moving at just the right speed for its altitude and mass so that it doesn't fall any further towards Earth. Gravity can't pull the Moon any closer, and the Moon's centrifugal force is not enough to enable

TERMINAL VELOCITY

An object dropped accelerates at 9.8m/s^2 until it reaches terminal velocity. It then continues to travel at the same speed until it hits the ground. It loses acceleration because the pressure of the air beneath the object and the drag as it passes through the air both work to slow it down. When the sum of the drag and the pressure from below (buoyancy) are equal to the force of gravity, acceleration stops. As gravity is a force which produces acceleration, cancelling its effect means the body stops accelerating and maintains its velocity (it doesn't mean it stops moving). Terminal velocity depends on the weight of the object (a measure of gravity acting on it), the shape of the object (which affects drag) and the medium it is falling through, such as air or water (which affects buoyancy).

A skydiver reaches a higher terminal velocity with a closed parachute than with an open one. When the parachute opens, it produces more drag so the equation changes and the skydiver's speed decreases. Eventually, a new terminal velocity is reached – a velocity which can produce a safe landing.

it to break free and whizz off into space. When we send a spaceship into space, beyond Earth's orbit, it must accelerate fast enough to overcome gravity. This is called escape velocity. It's effectively a kick in the teeth for gravity. If we want the spaceship to stay in orbit, it has to travel at just the right speed so that it doesn't fall to Earth and doesn't speed away. Satellites (including the Moon) are in freefall – they are subject to gravity, but as they are moving they don't actually fall.

Watching gravity between objects

Newton revealed his ideas about gravity in 1687 in perhaps the most important mathematical book ever

GRAVITY, MASS AND WEIGHT

Non-scientists tend to use mass and weight interchangeably, and that usually works perfectly well in everyday life. It doesn't matter whether you think of fruit you buy in terms of mass or weight – you'll get the same amount of fruit (as long as you're on Earth).

The mass of an object, measured in kilograms or pounds, is the same wherever the object is. But the weight of an object is a measure of how gravity acts on it and it varies in areas with different gravity. The Moon has only about one sixth the gravity of Earth, so instead of accelerating at a rate of $9.8 m/s^2$, an object dropped on the Moon accelerates at $1.63 m/s^2$. If the weight and mass of a box on Earth is 1kg (2.2lb), its mass would still be the same if you took it to the Moon, but it would only weigh 166g (0.36 lb).

published, *Philosophiæ Naturalis Principia Mathematica* (*Mathematical Principles of Natural Philosophy*). It was more than 100 years, though, before gravity between two small objects was experimentally demonstrated. Then, in 1796, English physicist Henry Cavendish used a torsion balance to measure the gravitational attraction between two objects. He suspended a dumbbell by a thread and used a heavy object to attract one of its weights, causing the dumbbell to turn towards it slightly, twisting the thread.

Henry Cavendish provided proof for Newton's theory that gravity operates between all objects with mass.

One, two, three, more

Inevitably, once we start thinking about gravity in space, we have to take account of more than just the first two bodies we thought of. The Moon is held in orbit by the gravity of Earth, but Earth is held in orbit around the Sun in exactly the same way. The Sun has an impact

WHY DOES A BALLOON GO UP?

Hot-air balloons and helium balloons both rise, apparently contradicting the theory of gravity. But they only go upwards if you release them in normal air. If you released them in a vacuum or in a room filled with hydrogen, they would fall like anything else. They rise because they are lighter than the surrounding air, so gravity exerts more force on the air around and above them than on the balloon filled with light gas. Helium is lighter than air, and hot air is less dense than cool air, so lighter than the same volume of cool air. As the air is pulled downwards around the balloon, the balloon moves upwards and air moves in underneath it to take its place.

on the Moon, making its orbit around Earth not perfectly circular. And there are many other bodies thrown into the mix which disconcert the orbits of the Moon, Earth and one another – they include planets and their moons, occasional comets and asteroids and even the slight impact of stars. There is no point at which an object is so far away that gravity just switches off for it. Instead, gravity gets ever weaker until it is infinitely small and its effects are no longer noticeable.

The result of many massive objects in a solar system is a great mix of gravitational forces acting in all directions, some having a very weak effect on distant objects, others having a strong effect on nearby objects. Calculating the exact orbits of planets and moons which are not even perfect spheres and are affected by the gravitational impact of many other bodies is currently impossible. Indeed, in 1887, the mathematicians Heinrich Bruns and Henri Poincaré showed that there is no general solution for the three-body problem – the patterns of gravitational attraction between three bodies - using familiar mathematical methods. Several 'families' of solutions are now known, but nothing that can be applied to all cases.

How does it work?

Newton admitted that although he had a theory about what gravity does, he couldn't say exactly how it does it. What *is* gravity?

> *'Gravity must be caused by an agent acting constantly according to certain laws. But whether this agent be material or immaterial, I have left to the consideration of my readers.'*
>
> Isaac Newton, 1687

The question of how gravity might work was left open for more than 200 years. Then the Austrian physicist Albert Einstein suggested in his general theory of relativity in 1916 that gravity is actually a distortion of the space-time continuum (see pages 222–3).

IF GRAVITY TURNED OFF . . .

There isn't a way for Earth's own gravity to 'turn off' as it's a necessary side effect of Earth existing at all. For gravity to disappear, Earth would have to disappear too, so we wouldn't be around to notice the effects.

The only way we could experience a sudden loss of gravity would be if we lost the Sun's gravity. Imagine the Sun suddenly explodes, so its gravitational pull disappears. If gravity travels at the speed of light, we wouldn't notice the Sun had gone for eight minutes, then we would see the explosion and Earth would lose the focus of its orbit at the same time. It wouldn't be a good day.

An artist's impression of two black holes colliding.

This did not entirely replace Newton's theory of gravity, which is still useful in practical applications but breaks down when considering very strong gravitational fields and very high speeds. Einstein's explanation of gravity works in situations where Newton's doesn't. His theory also provides a way gravity might work – by gravity waves that travel at the speed of light. The first evidence of gravity waves was found in 2015, produced by two black holes colliding 1.3 billion light years away.

Chapter 9
Phlogiston Theory

What makes things flammable? People once thought all flammable material contained a special substance.

Mysterious metals

THE IDEA

When something burns, it releases a fine substance or principle called 'phlogiston'. In some versions of the theory, phlogiston could even have negative mass, and its loss could increase the weight of a burned metal.

From general experience, we would expect matter to weigh less once it has been burned. Burning logs and paper are reduced to ash, which seems almost weightless. But there are exceptions: metals.

People probably noticed long ago that the mass of a metal increases when it is burned or heated if it forms a 'calx' (now known as an oxide). Paul Ech stated as much, around 1500, after he heated a mixture of silver and mercury and found it gained weight.

The Italian physician Giulio Cesare della Scala was on the right track when he suggested in 1557 that the weight gain of burning lead and rusting iron might be related, and both are caused by the absorption of some particles from the air. The French chemist Jean Rey also proposed that something from the air was added to the metal. He had no knowledge of the formation of compounds, or of the air as a combination of gases, so assumed it was some form of absorption rather than a chemical change – something like the way a cloth absorbs water on becoming wet.

Robert Boyle was the first significant British chemist.

Sadly, this line of reasoning did not prevail. The great English chemist Robert Boyle rejected the idea of air absorption and proposed instead that something from the flame itself was being added to the metal when burned: '[The] flame itself may be, as it were, incorporated with close and solid bodies, so as to increase their bulk and weight.' He weighed metal, sealed it in a glass container and exposed it to a flame. When he broke open the container and weighed the contents, he found the mass had increased. Even though he heard the hiss of air entering the container, he still dismissed the idea that air had been used up in some way, preferring to suggest that part of the flame ('igneous particles') had penetrated the glass walls of the vessel.

Towards a new model

In 1669, Johann Joachim Becher, a German physician and alchemist, found fault with the classical Greek scheme of four elements, earth, water, air and fire (see page 162), and even with a new scheme

LEONARDO GETS IT RIGHT

Far ahead of his time, the Italian polymath Leonardo da Vinci (1452–1519) recognized that air is necessary to sustain a fire. He wrote in his notebooks:

'The fire destroys without intermission the air which supports it and would produce a vacuum if other air did not come to supply it. So soon as the air is no longer in condition to sustain a flame, no earthly creature can live in it any more than can the flame.'

Because Leonardo wrote his notebooks in mirror-writing and did not publish them, his discovery had no wider impact.

introduced in the 16th century by the maverick Swiss physician Paracelsus (see box on page 135). Becher removed air and fire from the classical scheme and divided 'earth' into three types. They corresponded to Paracelsus's sulphur, mercury and salt, but Becher rejected Paracelsus's naming. One of his three types of earth he called *terra pinguis*, which he characterized as oily and sulphurous. He claimed it was present in everything that can be burned. It was, he said, released when substances burn, so was clearly a principle of inflammability.

Becher accepted the standard model of fire being a process of breaking down, so no pure elemental substance could burn – there had to be some component of *terra pinguis* to be released. That

For alchemists, fire was an essential element in transforming materials.

metals form a calx he attributed to the addition of some 'fire-matter', in much the same way as Boyle had suggested.

From *terra pinguis* to phlogiston

The principle of inflammability was picked up by Becher's devoted pupil Georg Stahl (1660-1734), who developed it into the theory of phlogiston. Phlogiston, he claimed, was a very fine (subtle) principle which combines with the four types of earth (counting water as fluid earth) and is present in all inflammable materials. Its name was derived from the Greek *phlogiston*, 'burning up'.

Stahl proposed that phlogiston was a universal component of flammable materials and was released on burning. The more phlogiston a substance contained, the less residue remained when the substance was burned. The basic principles of his thesis were:

- Phlogiston combines with matter, producing the quality of flammability.

- On its own (uncombined with matter), phlogiston cannot be detected by the human senses.
- Phlogiston constitutes the motive power (power of movement) of fire, and the movement produced is circular.
- It is the foundation of colour.

THREE ELEMENTS FROM ALCHEMY

Paracelsus was the first person to bring chemistry to bear on medicine, and as such was a hugely influential – though controversial – figure. He borrowed from alchemy the importance of three substances, mercury, sulphur and salt, with which he replaced the traditional elements and humours commonly believed at the time to be the root of all illness (see page 160).

According to Paracelsus, illnesses were produced by poisoning caused by one of these substances; the appropriate treatment depended on which was the cause in any particular case. He considered sulphur a combustible element, mercury a fluid and changeable element and salt a solid, permanent element. He demonstrated their interdependence by burning a piece of wood, claiming the fire itself was the work of the sulphur, the smoke the product of mercury and the ash the enduring residue, salt.

- A flammable substance loses phlogiston when it burns.
- Phlogiston cannot be destroyed, nor can it escape from the atmosphere (so it is perpetually recycled on Earth).
- Anything that cannot be burned contains no phlogiston. This also applies to the products of burning which can't be further reduced – they have lost all their phlogiston.
- Oil and fatty substances are rich in phlogiston.
- Heating a metal calx can restore the original metal by the addition of phlogiston.
- Both phlogiston and air are necessary to produce a flame.

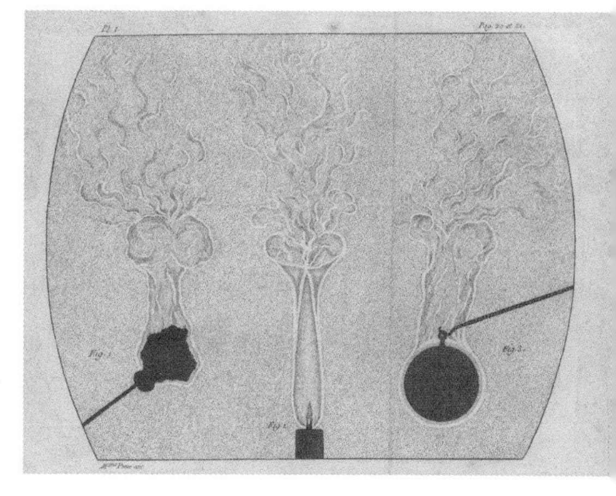

Jean-Paul Marat accurately illustrated the movement of hot gases around a flame or burning material (1780).

The tricky matter of weight

Stahl knew if he heated a calx with charcoal, he would end up with the original metal. So, for instance, heating zinc oxide with charcoal would produce zinc. He suggested that phlogiston in the charcoal combined with the calx, so reconstituting the original, phlogiston-bearing, metal.

It might seem strange that Stahl was aware that calces weigh more than the original metals, but still he was confident that the metal had lost phlogiston on burning. It seems he did not assume phlogiston to have any weight but to be, like light, entirely ethereal. He never speaks of it as a substance or type of matter, only as a 'principle'.

J.H. Pott, a student at the time, tried to clarify this, describing phlogiston as a principle like light or heat that cannot be separated from matter to investigate or measure on its own, but that it suffuses matter. Flames are visible when phlogiston is mixed with water. If phlogiston is considered to have no mass, then the explanation of Boyle and others for the addition of mass during the burning of metals is able to stand alongside the principle of losing phlogiston.

The theory persuaded many chemists of the day. As stated by Stahl, there was not really anything especially provocative in it. The more bizarre aspects of the theory were added by later phlogistonists.

Accruing absurdity

Among the refinements and developments of the phlogiston theory, the most famous is the suggestion that it imparts levity (lightness); that is, phlogiston has negative mass, so removing it *increases* the mass of the substance. This was suggested by Johann Juncker and was closely linked with, and perhaps an inevitable consequence of, his seeing phlogiston as a substance rather than a principle. If it were considered a substance, removing it should invariably reduce the mass of the burned material. The only way to avoid this conclusion was to give it a negative mass.

Not everyone found this a persuasive argument. Among its detractors was the English chemist Joseph Priestley. He was unwilling to accept that it conveyed levity, but acknowledged that heat and light can both affect the properties of matter and neither of those can be weighed. Priestley's recognition that there was a gap in the chemical knowledge of his day was by far the best way to consider the matter. An admission that 'we don't know' is always a better state of affairs than clinging to a nonsensical explanation with no foundation in evidence.

The end of phlogiston

The French chemist Guillaume-François Rouelle introduced the phlogiston theory to France where it immediately became popular. The version which took off was essentially that of Stahl, who maintained that

> ***'That phlogiston should communicate absolute levity to the bodies with which it is combined, is a supposition that I am not willing to have recourse to, though it would afford an easy solution of the difficulty.'***
> Joseph Priestley, 1774

Priestley experimented with gases, first isolating oxygen, nitrous oxide and several others.

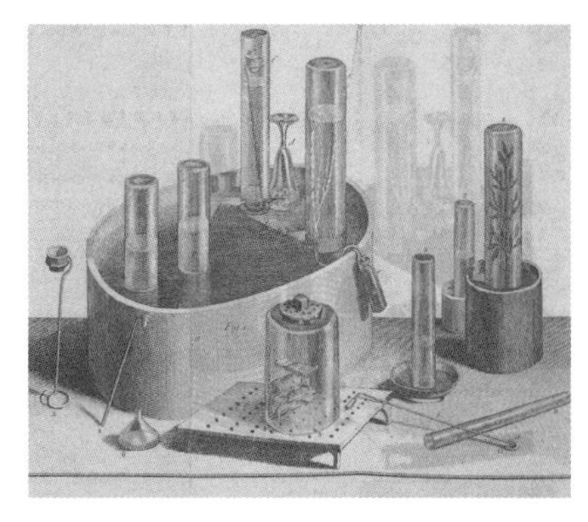

phlogiston is very 'subtle' and can't be discovered by analysis, but is present in all matter. It was one of Rouelle's students, Antoine-Laurent Lavoisier, who would eventually dismiss phlogiston forever.

In 1774, Joseph Priestley visited Lavoisier in Paris. He shared his finding that if he heated mercury calx he could collect a gas. A candle would burn well in the gas. This was how he discovered oxygen. Priestley considered this to be 'pure air', a kind of enhanced air that was good for both respiring animals and burning candles. In fact, he reported it was 'five or six times as good as common air' and a mouse sealed in a container of the gas would live about four times as long as a mouse sealed in a similar container of normal air. (This is all fair: air is 21 per cent oxygen.) Priestley, a fan of the phlogiston theory, thought this air was 'dephlogisticated air' – it contained no phlogiston, so could sustain a lot of burning activity before it became saturated with phlogiston and the candle would go out.

Lavoisier had already found that phosphorus and sulphur both gain weight when burned; if he heated lead calx a lot of gas was released, as Priestley had found with mercury calx. It was clear to Lavoisier that air is involved in burning in some way. After his discussions with Priestley, he repeated Priestley's experiments and came to the conclusion that air is not a simple substance. He proposed two ingredients: one supports breathing and combines with metals; the other causes asphyxiation and can't support burning. We now know that the first of these is oxygen.

A better answer

By 1777, Lavoisier had his own theory of combustion and respiration. He labelled part of the air 'eminently respirable', and explained that it combines with a metal or organic substance in combustion. Two years later, having discovered that most acids contained this breathable component of air, he named it oxygène, from the Greek meaning 'acid generator'.

Lavoisier was ready in 1783 to debunk phlogiston theory, calling phlogiston 'imaginary' and – as described – 'a veritable Proteus that changes its form every instant'. Priestley and another English chemist, Henry Cavendish, had discovered that if they set light to 'inflammable air' in the presence of ordinary air, a small amount of water formed. They explained this by saying that one gas is overlyphlogisticated water and the other dephlogisticated water, and that combining them produces properly phlogisticated water. Lavoisier's simpler (and correct) explanation was that water is composed of the two gases. As combustion consists of adding oxygen, oxygen is one half of water and inflammable air (which he named hydrogen) is the other half. The entire reaction was explicable without resorting to phlogiston. Indeed, Lavoisier could explain respiration and combustion with reference only to oxygen, with no need for the mysterious, undetectable phlogiston.

He was well aware that scientists committed to phlogiston theory would be reluctant to give it up and be prone to 'adopt new ideas only with difficulty'. However, by 1791 he could report: 'All young chemists adopt the theory.' It was as well they did so quickly enough to gratify Lavoisier – he was executed by guillotine in 1794, a victim of the French Revolution.

Lavoisier's experiments with gases and respiration were recorded by his wife.

Why would the coast of South America fit so well next to the coast of West Africa?

THE IDEA

The Earth's crust – the top layer, carrying the oceans and continents – is broken into a series of tectonic plates which are slowly moving, carried by currents in the molten rock (magma) of the Earth's mantle. The continents have reconfigured many times over Earth's history.

Jigsaw pieces

Not long after the discovery and first mapping of the Americas by Europeans, people began to notice that the coast of Central and South America fits quite neatly against the coast of West Africa. The Dutch mapmaker Abraham Ortelius suggested in 1596 that the Americas had been 'torn away from Europe and Africa . . . by earthquakes and floods'.

The English philosopher and scientist Francis Bacon commented on the fit in 1620, but could suggest

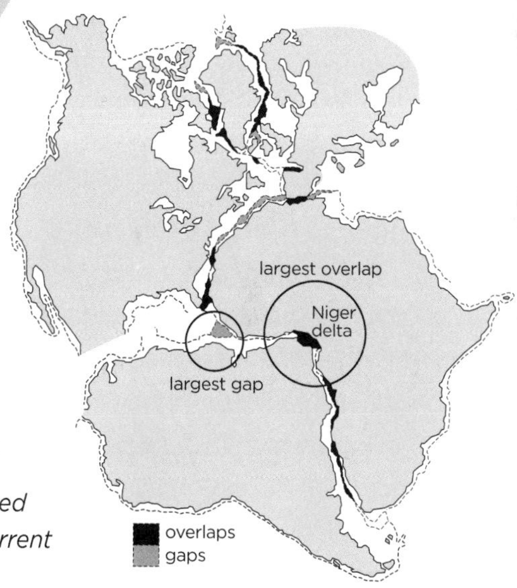

The best fit between coastlines is achieved if the outline 1,000m (3,280ft) below current sea level is used for comparison.

no explanation. The Church taught that the Earth had been made perfect and unchanging in God's act of Creation, so the idea that bits of it might be shuffling around was contentious. But the shape of the coasts of the two continents made it difficult to ignore the possibility.

By 1750, the French naturalist Georges-Louis Leclerc was willing to propose that the continents had once been joined, but he could not offer a mechanism for their moving apart.

Rocks and animals, here and there

Other evidence soon emerged to support the idea of shifting continents, though. The German explorer and naturalist Alexander von Humboldt found similar rocks in Brazil and the Congo. He hypothesized that the continents were once joined but the connecting land bridge was destroyed by a great tidal wave which left the Atlantic Ocean in its place.

The mystery deepened with the discovery that there were living animals and plants on the separated continents which resembled one another. The fossil record also suggested a close correspondence between animals inhabiting the landmasses in the past. Fossils of animals which could not possibly have crossed the Atlantic on their own were found on each side of the ocean.

Similarities between the creatures and rocks of other separated lands began to emerge, too. The conclusion that the land had moved became irresistible. But was it the result of catastrophic events or of more gradual processes, or perhaps both? Flooding is mentioned in the Bible and in many other religions and myths around the world, so that was the less controversial proposal.

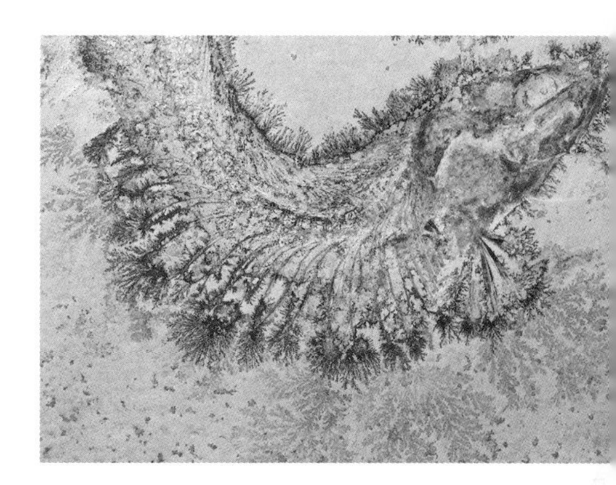

This fossil fish, found inland in Germany, is evidence that rock which was once beneath the sea is now dry land.

One big chunk?

Then, in 1885, Austrian geologist Eduard Suess proposed that once all the landmasses had been clumped together in two giant continents which he named Gondwanaland and Laurasia. Suess's work in paleobiology led him to the discovery that an extinct fern-like tree called *Glossopteris* grew in India, South Africa, South America and Australia. He reasoned that the lands must once have been joined or close enough for seeds to be carried between them. He couldn't explain how they had broken up nor how the parts had moved to the current location of the continental landmasses.

The model of Earth's history at the time had a hot world slowly cooling. As it cooled it shrank, so the surface wrinkled. Suess assumed that the shrinking produced movement. It would also produce a rise in sea levels, which could submerge land bridges in a less catastrophic way than giant tidal waves. But the answer was not a shrinking planet.

Shifting places

In 1912, German meteorologist and geophysicist Alfred Wegener proposed his theory of continental drift. It built on previous speculation about possibly shifting continents to propose a proper scheme of what had happened and continues to happen.

Drawn like others before him to the odd congruence between continental coastlines, Wegener began to explore the possibilities. He was soon convinced that the landmasses which seem to fit together like the pieces of a jigsaw had once been adjacent. Several types of evidence drew him to his final conclusion: the landmasses had once all been joined in a single, vast supercontinent which he called Pangaea (meaning 'all land').

Wegener was not persuaded by the idea of land bridges. He pointed out that the rock that makes up the continental landmasses is predominantly granite, while the seabed is made of denser volcanic basalt. He suggested that the continents effectively float like icebergs on a sea of basalt. If there had been a land bridge and it had been pushed below the sea, it would eventually rise again when any force pushing it down was removed. Wegener pointed to the way the lands of the northern hemisphere were slowly rising after being weighed down by ice during the most recent ice age.

> ***'Doesn't the east coast of South America fit exactly against the west coast of Africa, as if they had once been joined? This is an idea I'll have to pursue.'***
> Alfred Wegener, 1910

Wegener also noted that the geological compositions of South America and Africa are contiguous. Areas of ancient rock called cratons, 2,000 million years old, had apparently been torn apart and were stranded on either side of the Atlantic Ocean. Furthermore, he noticed a massive belt of mountains which had once straddled the northern Atlantic Ocean, running from Greenland through the UK, Canada and northern Europe. These form the Caledonian mountains, now thought to have been created 400–500 million years

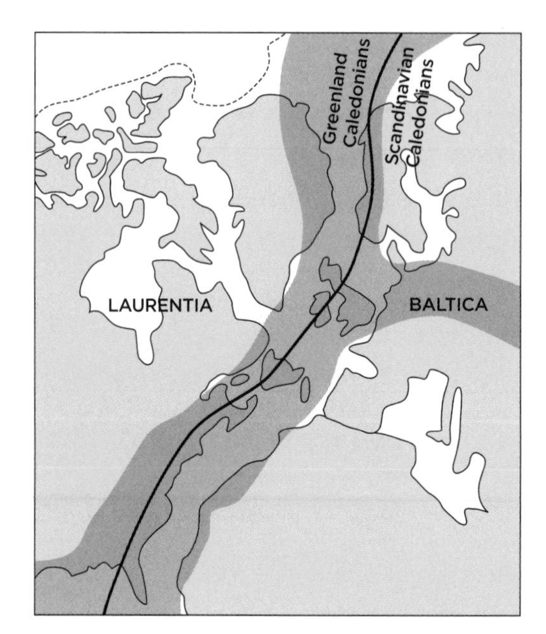

ago when landmasses came together in the formation of Pangaea. Wegener recognized that they were a single band, prised apart by continental drift.

Passengers

Rocks carry stuff with them as they move. These bits of geological debris and fossils can help to reveal their past journeys.

Glacial deposits are chunks of rock and soil carried by slow-moving glaciers and dropped en route, showing the path of the glacier's movement. Wegener described glacial deposits formed around 280 million years ago and found in India, Australia, Antarctica and South America. They date from a time when ice extended further from the poles than it does now; if the continents had been in their current positions the ice would have had to stretch as far as the north of India – that would be serious global glaciation!

Suess had already noted that fossils of the tree *Glossopteris* are found on both sides of the Atlantic. More exciting organisms share the same split range, including the large reptile *Mesosaurus*. This crocodile-like creature is found only in a band across South America and South Africa. How could such an animal have crossed the Atlantic Ocean? Wegener reasonably concluded it could not. Losing the Atlantic Ocean was more credible than either a crocodile crossing thousands of kilometres of open sea or the same animal evolving independently on each side of the ocean.

'Utter, damned rot'

Wegener's theory sounds entirely reasonable and compelling now. But people were still inclined to cling to the biblical account of an unchanging world created at a stroke by God, so the idea that the vast, solid landmasses moved around was provocative to say the least. It didn't help that Wegener had no good way of explaining how or why the continents might move.

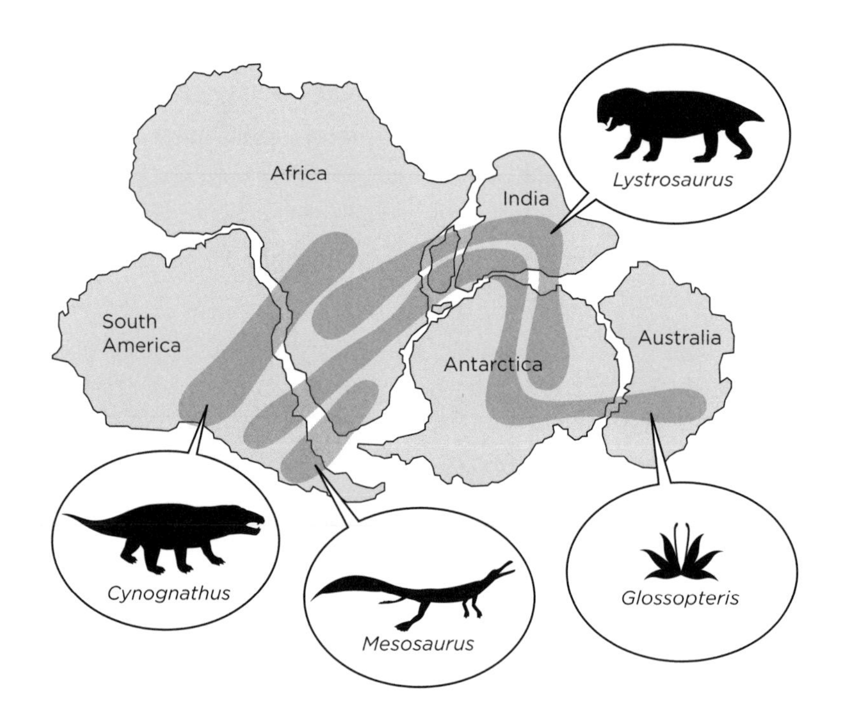

He first published his ideas on what he called continental drift in 1912, and more fully in 1915 in *The Origin of Continents and Oceans*. In the 1922 edition of his book he introduced the supercontinent Pangaea, suggesting that it began to break up around 200 million years ago. His work was met with hostility and sometimes ridicule. The president of the American Philosophical Society called it 'Utter, damned rot!' Others attacked Wegener personally, claiming that as a meteorologist he was invading territory about which he knew nothing. Wegener remained sanguine in the face of opposition: 'It is probable the complete solution of the problem of the forces will be a long time coming. The Newton of drift theory has not yet appeared.' (1929)

> ***'It is just as if we were to refit the torn pieces of a newspaper by matching their edges and then check whether the lines of print ran smoothly across. If they do, there is nothing left but to conclude that the pieces were in fact joined in this way.'***
> Alfred Wegener, 1915

Ironically, the English geologist Arthur Holmes had already suggested in 1928 that convection currents within the Earth's mantle might be the means by which continents moved. His suggestion went unremarked, though later tectonic plate theory would bear it out.

Mountains, plains and trenches – under the sea

Wegener was right; it took until the 1960s and 1970s for a good explanation to emerge. In between were decades of confusion. Articles published in Europe often went unnoticed in the USA, and the political and economic chaos of the 1920s, 1930s and 1940s didn't help. When Wegener died in 1930, geologists were divided between 'drifters' or 'mobilists' who supported the theory of continental drift and 'fixists', who were convinced that the continents do not move.

Then, in 1962, American geologist Harry Hess discovered something unexpected. Using new sonar techniques to map the seafloor of the North Pacific, he found undersea ridges like mountain ranges, running through the middle of the ocean and making the sea above shallower than elsewhere. Another surprise was that the deepest parts of the ocean are close to the continental shelf, the area where the continents rise up out of the sea. Deep trenches run around the margins of the continents, plunging as far as 11km (7 miles) in the case of the Mariana Trench off the coast of Japan.

At a subduction zone, an oceanic plate goes beneath the edge of the continental plate and melts back into the mantle.

The majority of the seabed, called the abyssal plain, lies between the undersea mountains and the trenches and is only around 4km (2.5 miles) below sea level.

Hess suggested that the rock of the seabed is created in the middle of the ocean and oozes up through cracks in the Earth's crust as thick, hot, molten basalt. This piles up to form the mid-ocean ridges. The new rock moves gradually outwards from the centre, spreading to both sides. Hess believed the trenches near the coast were the sites of old seabed being disposed of in some way. This tackled a problem raised by fixists – that if the oceans kept producing new rock the Earth should be growing in size, but there was no evidence to suggest it was.

Many of the ideas put forward in Hess's explanation had been mooted earlier, but now the world was ready to pay attention. Hess suggested that convection currents within the mantle dragged rock from the middle of the oceans to the edges, where it was consumed. Like others before him, he assumed that the old crust was stuck on to the continental landmasses, making them thicker. He was wrong.

Magnetism and magma

The first piece of scientific evidence to support Hess's theory of seafloor spreading fell into place in 1963.

Two British geologists, Frederick Vine and Drummond Matthews, showed that the age of the rock and its magnetic orientation were symmetrical either side of the mid-ocean ridges. The magnetic striping of rock on the ocean floor had been discovered in 1961 (see box below), but not related to the mid-ocean ridges.

If they looked at rock, say, 100km (60 miles) either side of a ridge, Vine and Matthews found that the age of the rock and the record of

EARTH'S SWITCHING MAGNETIC FIELD

The Earth's magnetic poles switch periodically, with the north magnetic pole becoming south and vice versa. A record of the polarity is preserved in some rocks because the seabed is rich in magnetite, a strongly magnetic mineral. As the rocks cool, the magnetite lines up with the Earth's magnetic field and is locked into place, preserving a record of the polarity at the time.

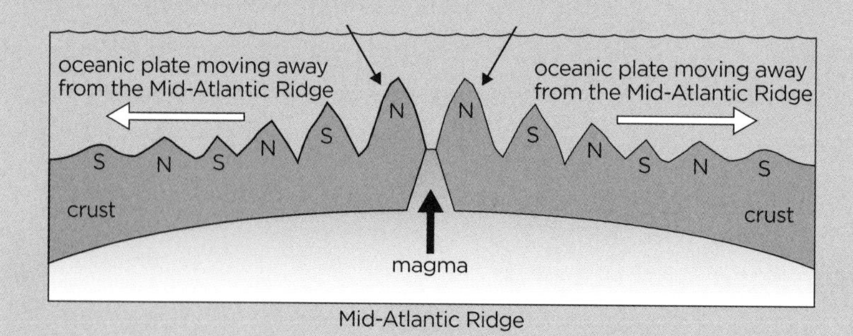

magnetic orientation frozen into it were identical on both sides. This strongly supported the idea that the rock was created at the same time and pushed out sideways from the ridge in both directions equally. Canadian geophysicist Lawrence Morley came to the same conclusion independently, and this model is now known as the Vine-Matthews-Morley hypothesis.

Recycling on a grand scale

But where was all the spare rock going at the edges of the continents? Early developments in seismic imaging showed oceanic crust can disappear into the deep trenches off the coast. The 'spare' rock pushed out of the way by the new rock oozing from the mid-ocean ridges could be reabsorbed into the mantle.

The seafloor spreading hypothesis provided a neat explanation which married magnetic striping, rock production at the mid-ocean ridges and a potential mechanism for moving new rock from the centre of the ocean to the landmasses. But it didn't explain where the oceans came from in the first place or why Africa and South America might have split apart. A bit more was needed.

Zoning in on earthquakes

In the 1920s, scientists noticed earthquakes don't happen randomly around the globe but tend to cluster along lines. It turns out that

Earthquakes such as this in Alaska, in 1964, are caused by pressure built up between tectonic plates grinding against each other suddenly being released as they move.

these lines generally fall along the mid-ocean ridges and in areas parallel to the trenches off the continental coast. Mapping the frequency of earthquakes produced an image of the globe broken into discrete pieces – rather like a plate which has been smashed and glued back together.

Faults and fault lines

The boundaries between the pieces became known as 'faults'. Two types were recognized first: one at the mid-ocean ridges, where new crustal rock leaked out and spread from beneath, and one near the margins of the continental landmasses where old rock from the seafloor was somehow reabsorbed and recycled. At the sites of spreading, where these were close to the surface (as in Iceland) volcanic eruptions were common. At the sites of recycling, both earthquakes and volcanic eruptions were common. In 1965, the Canadian geophysicist John Tuzo-Wilson described a third kind of fault line, where moving slabs of the Earth's crust grate against each other while moving sideways in opposite directions. These were characterized by earthquakes.

From continental drift to plate tectonics

In 1965, English geophysicist Sir Edward Bullard revisited Wegener's theory of continental drift, showing how the continents could best be

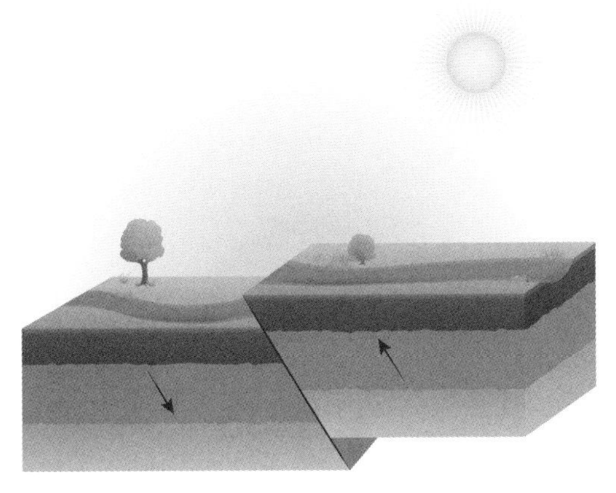

At a transverse fault, plates grind against each other causing earthquakes.

fitted together (called 'Bullard's fit', see picture below). Over the next two years, plate tectonic theory fell into place.

In 1966, Tuzo-Wilson proposed that the movement of the landmasses was just one step in an endless cycle. The 'Wilson Cycle' outlines a sequence in which the landmasses repeatedly group together to form a supercontinent and then break apart as oceans open up between them. The full cycle takes 300–500 million years to complete.

In 1967, W. Jason Morgan suggested how this might work. The Earth's crust is divided into 12 'plates' which move relative to one

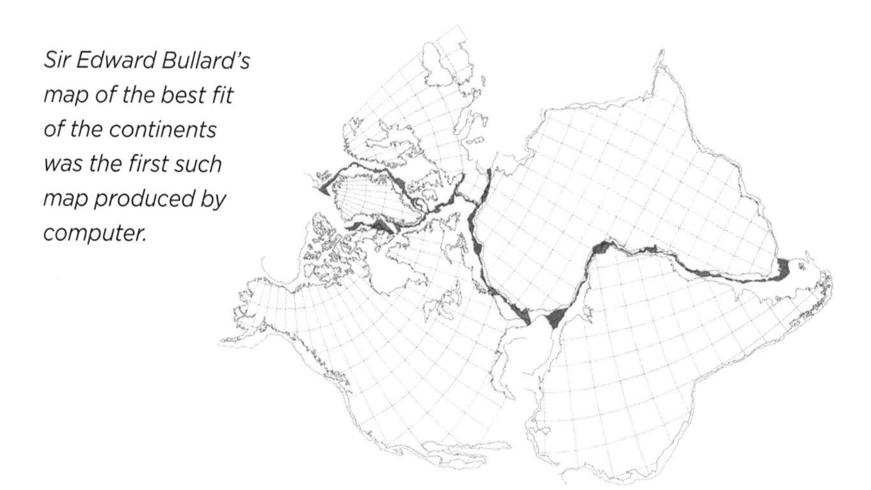

Sir Edward Bullard's map of the best fit of the continents was the first such map produced by computer.

another, effectively jostling their way around the planet. These crustal plates are carried by convection currents within the magma of the Earth's mantle. New rock wells up at the mid-ocean ridges where the plates are pulling apart and old rock descends beneath the continental plate at the ocean's margins where the plate carrying ocean sinks below the edge of the lighter plate carrying land. Where two plates are forced against each other, mountains are pushed up. Where the plates grind alongside each other, earthquakes wrack the land or seabed. Tectonic plate theory, at last, accounts for all the various types of geological activity we witness.

A tectonic crack in Iceland.

Chapter 11

Humoral Theory

For around 2,000 years, people believed that human temperament and illness were dependent on four 'humours'.

Being sanguine

If someone is said to be sanguine about something, it means they are optimistic, confident and positive. The origins of the term, meaning 'full-blooded' or 'full of blood', lie with the Greek medical writer Hippocrates in the 4th–5th century BC.

THE IDEA

The balance of four 'humours' or bodily fluids determine a person's character and state of health. Illness can be treated by rebalancing the humours.

Hippocrates identified four bodily fluids, blood, yellow bile, black bile and phlegm, which he associated with the four classical elements. When they were out of balance, the result was illness or pain. Humoral theory became the bedrock of medical thinking. It's possible that the notion was old even in Hippocrates' time, maybe originating in Mesopotamia or Ancient Egypt around 4,000 years ago.

The Roman physician Galen developed the humours further, adding a psychological aspect. He found an excess of one humour led to personal

Hippocrates is sometimes considered the father of Western medicine.

The melancholic figure would today be called depressed.

temperaments he described as sanguine (excess blood), choleric or aggressive/angry (yellow bile), melancholic/depressive (black bile) and phlegmatic/calm (phlegm). The four humours were also associated with the properties of the four elements – fire, air, water and earth – with the characteristics, respectively, of hot/dry, hot/wet, cold/wet, cold/dry. The ideal temperament was perfectly balanced between the various humours and elements.

In sickness and in health

According to humoral theory, each individual has a personal balance of humours which determines his or her temperament. If this balance is disturbed, the result is illness. Galen taught that illness was corrected by rebalancing the humours. Sometimes the cure was quite innocuous or even pleasant – the heat and moisture of a hot bath, for example – but often it was not. Many unfortunate patients were treated with bloodletting or emetics (drugs to induce vomiting) to rid their bodies of excess blood or yellow bile. Bloodletting was often used in the most inappropriate cases, with a person already weakened by disease and even blood loss being induced to lose more blood through cutting, cupping (applying special cups to the skin to create suction) or leeches.

ALL IN THE BLOOD

If a vial of blood is left to settle overnight, it produces four layers: black clot-like material at the bottom, a layer of red blood cells, a whitish layer and then yellow-tinted serum at the top. It is easy to see how this could have been interpreted by early physicians as the separation of the four humours.

As treatment required restoring balance, it could involve increasing or decreasing the presence of one or more humours. Increasing humours was usually accomplished through diet or medications. As the same properties were also attributed to foods, diet could be used to help rebalance the humours. For instance, a person diagnosed with melancholy was deemed to have an excess of black bile, with the qualities of being cold and dry. Hot and wet foods would be best suited to treating melancholy as they would boost the

opposing qualities. Black bile could be expelled from the body with a purgative (leading to diarrhoea).

Inside out

Although humoral theory began as a medical theory accounting for physical health, it expanded during the Middle Ages to have far wider implications and applications. If a person's natural individual humoral composition was unbalanced, their natural temperament could be dysfunctional and hard to remedy. In 1621, Robert Burton wrote a long treatise on melancholy. He defined it as: 'a habit, a serious ailment, a settled humour . . . not errant, but fixed: and as it was long increasing, so, now being (pleasant or painful) grown to a habit, it will hardly be removed.'

Humorous faces

The effect of an excess of a particular humour was considered to be externally visible. So a person's melancholic temperament (as opposed to a person suffering a temporary bout of melancholy) would be written on their face and, indeed, their entire physique.

Knowledge of the traits associated with different humoral types was so widespread that poets and artists used them as shorthand for conveying a character's temperament. In Chaucer's *Canterbury Tales*, a series of pilgrims each tells a story on their journey to Canterbury.

The physical descriptions of the pilgrims are not random; some of their features would have alerted Chaucer's audience to particular personality types. The Reeve, for instance, is described as slim with thin legs, which would indicate a choleric nature – quick to anger, sharp-witted and probably lustful. The audience could anticipate the type of tale a character was likely to tell by observing the clues in their personal appearance.

A more 'scientific' approach was promoted by the Swiss physiognomist Johann Lavater in the 18th century. He had been influenced by a 16th-century book by Giambattista della Porta, who illustrated his text on physiognomy with pictures of people who looked like animals (and were deemed to share characteristics with those animals).

Lavater returned to the humours as a source of insight into individual personalities. He showed the characteristic facial features, as he understood them, of the temperaments dominated by each humour. This was, he claimed, a starting point to reading someone's personality.

The idea that there is a parallel between external appearance and internal qualities is hard to dispel. Western literature of the 19th century is full of characters whose appearance is a shorthand for their inner life; they are particularly prevalent in the novels of Charles Dickens, who never wastes an opportunity to make a virtuous woman beautiful or a villain ugly. It is still with us: when did you last encounter an obese, ugly character in a film or novel who was characterized as generous, kind and intelligent? But in real life you might well know such people.

The beginning of the end

The first medical book with a more modern approach was *De Humani Corporis Fabrica* (*On the fabric of the human body*), written in 1543 by Belgian anatomist Andreas Vesalius. It presented a serious challenge to the medical tradition rooted in the work of Galen, and this was very unpopular. Then, in 1628, English physician and anatomist William Harvey published *Exercitatio Anatomica de Motu Cordis et Sanguinis in Animalibus* (*An Anatomical Study of the Motion of the Heart and of the Blood in Animals*) in which he explained the motion of the heart when pumping blood around the body. Further explanations appeared, overturning some of the long-held beliefs that had originated with Hippocrates and Galen.

The criminal, Bill Sikes – one of Dickens' famous villains.

Giambattista della Porta illustrated his text with pictures of people who looked like animals.

But people were unwilling to give up the familiar and secure model of a body controlled by humours and for nearly 300 years the theory lived on. Patients still received – and asked for – emetics, purges and bloodletting sessions. These treatments possibly had some psychosomatic effect; also, people who would have got better anyway attributed their recovery to the humour-based treatments they received. But eventually medical science moved away from the humoral model entirely.

Dead but not gone

Although it would be difficult to find anyone now who believes in humoral theory, echoes and remnants persist. The words 'sanguine' and 'phlegmatic' are still used metaphorically to denote personality types, though with no belief that the person genuinely has an excess of blood or phlegm.

The notion that a person's character can be read in their appearance is no longer based on humoral theory but it still has

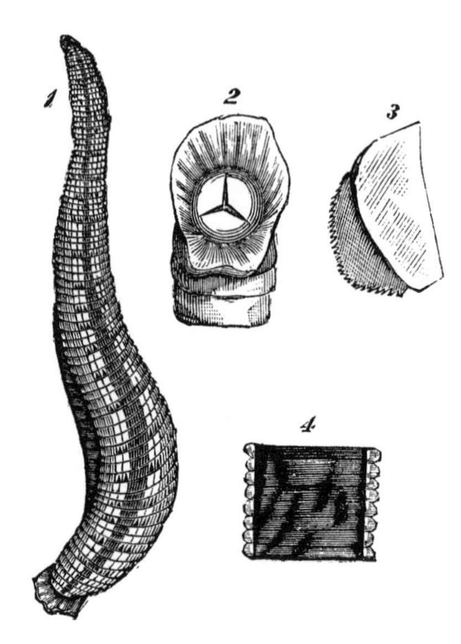

Leeches provided a quick, easy and non-messy way of bleeding patients.

currency. Most of us, whether or not we believe physical characteristics are a clue to personality, make quick judgments based on appearance and first impressions. And we still speak of people who are 'well-balanced'.

Measuring personality

Today, most psychologists would agree that personality is determined by a mixture of genetic factors and the results of experience – a combination of nature and nurture. The 'humoral temperaments', then, correspond to the inherited parts of personality today.

The German-born psychologist Hans Eysenck (1916–97) devised a system of defining temperament based on two axes, plotting: extraversion/introversion (E) and neuroticism/stability (N). This gave four categories, just as the humoral theory had done:

- Stable extraverts (sanguine-like: outgoing, lively, talkative, carefree, easy-going)

- Unstable extraverts (choleric-like: restless, irresponsible, excitable, impulsive)
- Stable introverts (phlegmatic-like: even-tempered, thoughtful, reliable, controlled, calm, careful)
- Unstable introverts (melancholic-like: reserved, pessimistic, moody, anxious).

Eysenck eventually found these categories inadequate and added another pairing, psychoticism/socialization, which put people on a spectrum from aggressiveness and non-conformity to compliance and consideration. His initial axes map remarkably well to those of the humours, suggesting not that humoral theory was correct but that it was used to express and explain aspects of temperament that either genuinely exist or are commonly perceived across cultures and time periods.

On balance

There have been many medical systems around the world that have focused on maintaining a balance in the body, either of substances or of energy. The modern biological concept of homeostasis is, in effect, a matter of balance – keeping the body in equilibrium. Humoral theory was wrong in that it relied on balancing something that doesn't exist, but it was right in principle – the health of the body does depend on maintaining the right levels of different chemicals.

Chapter 12

Simulation Theory

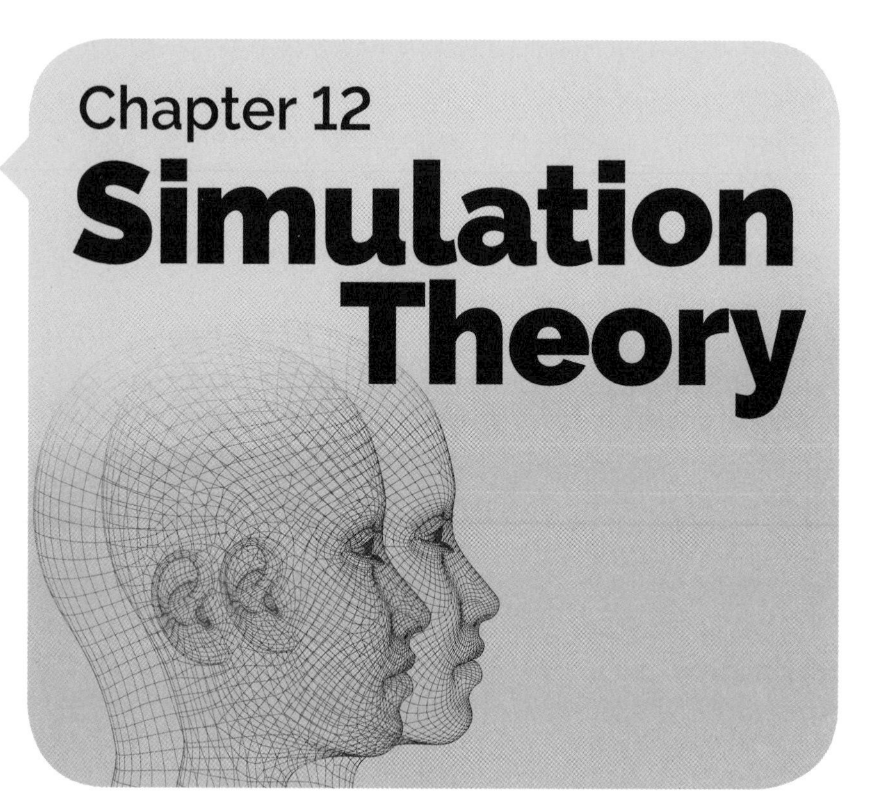

Descartes famously claimed: 'I think, therefore I am.' Could he have been entirely wrong? Could he, perhaps, think but not be real at all?

THE IDEA

The world around us is not real and nor are we; we are 'living' inside a computer simulation generated by a much more advanced civilization – more specifically, our own future, post-human civilization.

Is this idea for real?

The notion that we might not exist at all but be purely part of a computer simulation put together by some far more intelligent beings sounds far-fetched. But an increasing number of scientists and philosophers are prepared to entertain the idea. Indeed, the Isaac Asimov Memorial Debate at the American Museum of Natural History in 2016 addressed this very question.

> ***'I think the likelihood [that we are part of a simulation] may be very high . . . it is easy for me to imagine that everything in our lives is just a creation of some other entity for their entertainment.'***
>
> Neil deGrasse Tyson, director of the Hayden Planetarium, New York, 2016

What's real?

For millennia, philosophers have addressed the question of how we can be certain what we see and experience is real. The answer seems to be that we can't be sure. Even Descartes suggested there could be a demon constructing the world and fooling him into believing it exists.

The American philosopher Gilbert Harman updated the scenario in the 1970s, suggesting that if a brain were connected to a computer and fed the information to have experiences, these experiences would be indistinguishable from reality. This part at least seems to be borne out by discoveries in modern neuroscience. It is possible to stimulate areas of the brain to produce experiences which have no correlation with external reality, from feeling pain in an uninjured part of the body to seeing light flashes when there are none. But these scenarios do at least require the brain to be real. Simulation

theory goes beyond that to suggest that none of all this is real.

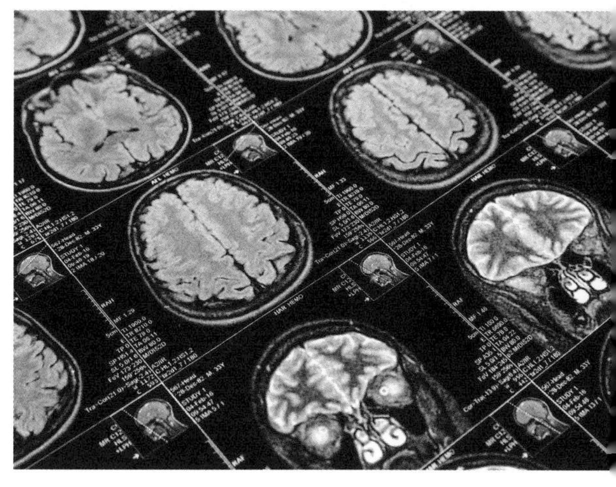

Modern neuroscience can access areas of the brain that were a mystery to our ancestors.

The world as simulation

The Swedish philosopher Nick Bostrom set the ball rolling with a paper published in 2003. He began by proposing that at least one of these propositions is true:

1 Humans will probably become extinct before reaching a 'post-human' stage.

2 Any post-human civilization is extremely unlikely to run multiple simulations of their evolutionary history.

3 We are almost certainly living in a computer simulation.

The argument suggests that if post-humans do run simulations, there are likely to be so many simulations by contrast with a single (past) reality, that we are far more likely to inhabit one of the simulations than the reality. And if we want to believe we will have descendants far in the future who might run such simulations, we have no right to assume we are living in reality. It's a bit hard to get your head around the idea that these post-people 'in the future' already exist outside the simulation we are in, as if the future is a sort of outdoor 'now'.

For us to be existing in a simulation it's necessary to accept that there can be nearly endless computer power and computers can create consciousness. The future ability to produce plausible virtual reality (VR) is pretty much a given, as we're quite good at this already. It's clearly not very hard and should be easily within the capabilities of intelligent post-humans. The bit about computers'

> ***'One thing that later generations might do with their super-powerful computers is run detailed simulations of their forebears or of people like their forebears. . . . Then it could be the case that the vast majority of minds like ours do not belong to the original race but rather to people simulated by the advanced descendants of an original race.'***
> Nick Bostrom, 2003

ability to create consciousness remains contentious, though many philosophers and computer experts believe it could be possible. Bostrom accepts that post-humans might not be interested in running ancestor-simulations, but if they were, the number of levels of simulation and simulated individuals would proliferate rapidly. There would be nothing to stop the simulated ancestors beginning their own simulations once their technology becomes sufficiently

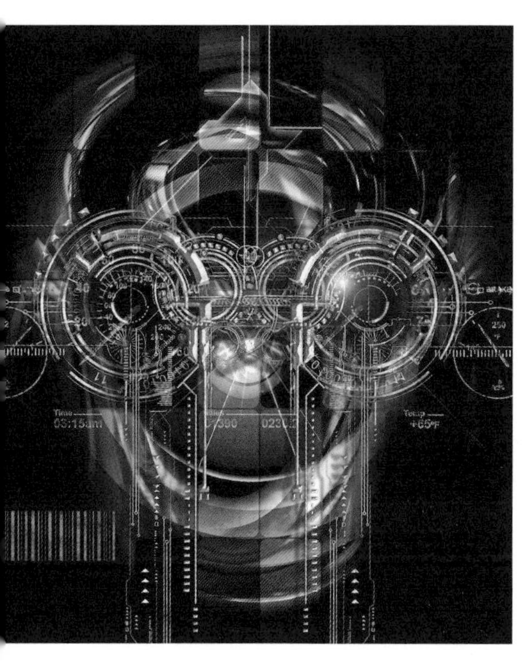

advanced. This means we might already be several layers of simulation down.

Be good – you don't know who's watching

Bostrom further suggests that if you realize you exist in a simulation, you might well start trying to second-guess what the simulator wants to see and begin acting in a way that will bring you favour. A form of afterlife becomes possible, and the random contingency of life might be open to influence. Perhaps the simulators favour simulated beings who act in a certain way, or perhaps they will intervene directly? It all begins to sound a bit like having a god.

People might behave in a morally good way in the hope of pleasing the simulators. Or they might just behave in a 'look at me' way, trying to be one of the more interesting simulations. The ultimate simulators will probably have realized that they can't be certain they are not simulations either, so there might be a chain of virtue (or a chain of showing-off) extending through the simulated levels.

Bostrom's conclusion is that 'unless we are now living in a simulation, our descendants will almost certainly never run an ancestor-simulation'.

Why think that?

The fact that the laws of physics and maths seem to work in new, unknown situations is one reason for believing that the universe around us might be a simulation. It endorses the view that a superior intelligence has produced a consistent and coherent simulation and we are, as it were, discovering the programming rules. But it could just as well be that a real universe follows physical laws which we have correctly discovered, which is why they are manifest in many situations. We have also found that everything breaks down into tiny particles, just as a computer image breaks down into pixels. (Of course, if reality really is broken into particles, then anything we produce must also be broken into particles at some scale, and whether it's pixels or photons the argument would be the same.)

But how would we find out whether we are a simulation or the real thing? Zohreh Davoudi, a physicist at MIT, suggests that the programmers might have tried to cut corners to save resources – though if they're short of resources, making millions of simulations of their ancestors doesn't seem a sensible use of those they have.

As Bostrom proposed that our post-human descendants might use nanotechnology to convert entire planets into massive computers, the limit on computing power would be the number of planets they are prepared to kidnap and destroy.

We might look for limits to the universe, as a simulation would not be infinite. But the universe might be finite anyway. Or we could find evidence that space-time is not continuous. Again, although this discovery wouldn't support our current theories of what the universe

> ***'Is it logically possible that we are in a simulation? Yes. Are we probably in a simulation? I would say no.'***
> Max Tegmark, professor of physics, MIT, 2016

Not just our consciousness, but everything we observe and experience must also be simulated.

is like, it wouldn't prove we live in a simulation – our theories about the universe could just be wrong. It wouldn't be possible to prove the universe is real but we could, arguably, prove it's not.

Does it matter?

If you decide you are simulated, you probably won't live your life very differently. However, some proponents of the theory suggest you should try to be super-interesting so the game-player keeps you going and maybe even gives you extra resources to play with. The reward of being elevated to the next level might make you behave better – but it's hard to second-guess the moral code of future post-people. Or you could become less concerned with morality and the good of other people – if we're all simulated, does it matter? (Of course, it does matter if the simulated beings have consciousness and feelings, as we seem to.) It's more likely we will carry on exactly as before, though there's a danger that the players simulating us might get bored, or want the computer for a new project. Then they might just click 'Exit' and it's all over for us.

We want to break free . . .

However crazy the idea might sound, two billionaires in the USA are funding scientists to try to find a way to break out of the simulation into reality. It's not clear how they intend to do it – or whether a simulated being can break out anyway. Perhaps – if it's not all nonsense – they will just end up getting us turned off, or we will step out of simulation into non-existence. That wouldn't be good either!

> ***'Maybe we should be hopeful that this is a simulation. Either we're going to create simulations that are indistinguishable from reality or civilization will cease to exist. Those are the two options.'***
> Elon Musk, 2016

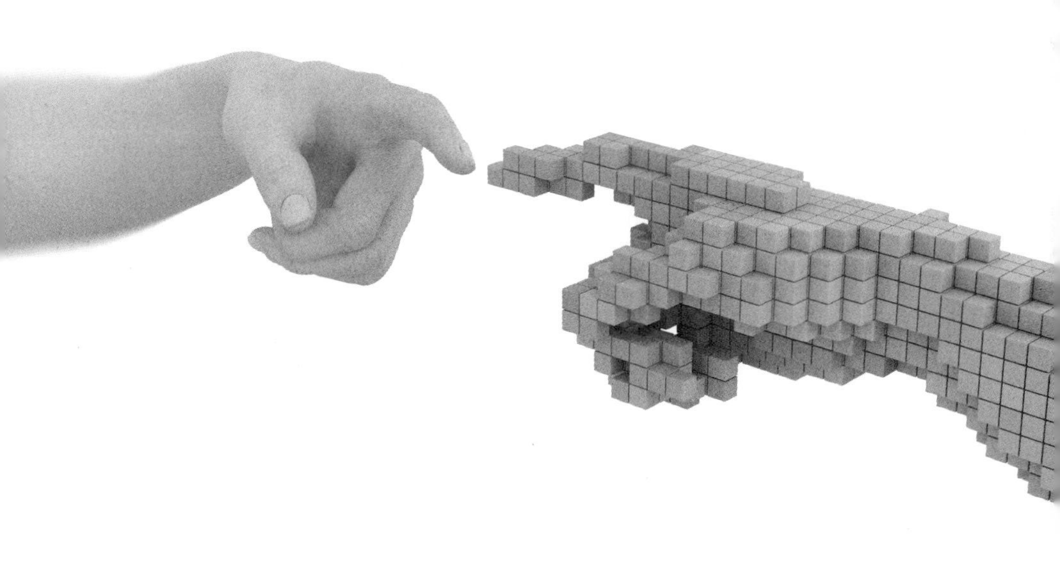

Chapter 13

Giant Impact Theory

Many religions and cultures have had myths about the Moon. So how might Earth have acquired its only natural satellite?

THE IDEA
The Moon formed after another planet crashed into the Earth 4.5 billion years ago.

Early ideas

Humans have wondered about the origins of the Moon for millennia. Early on, they made up stories to explain it, but these were not based on any observed evidence. As people turned towards scientific thinking, their ideas about the Moon's origins began to fit with what they knew about the solar system. There have been three principal theories: it was a body captured as it wandered past; it formed alongside the Earth; or it formed from matter ejected by the Earth.

The captive theory discarded

One possibility is that the Moon formed as an asteroid elsewhere in the solar system and was captured by Earth's gravity and dragged into orbit. This is thought to be how Mars gained its two moons, Phobos and Deimos. But it's unlikely to be the case with the Moon, as its chemical composition is too similar to that of Earth for it to be likely to have formed independently elsewhere. It would also have had to slow down by just the right amount at just the right point to be captured by (rather than crash into) the Earth.

Spitting out rocks

In the 1880s, George Darwin, son of the more famous Charles, suggested that the early Earth, spinning more quickly than now, had thrown out a great globule of its molten surface which hardened into the Moon. One idea was that the Pacific Ocean might have been gouged out by the removal of this big blob of Earth. This has since been debunked: the composition of the rock under the ocean doesn't match that of the Moon and the ocean is too young to have been created by the loss of the Moon.

Until recently, modern scientists considered it unlikely that the Earth could have been spinning fast enough to hurl out the material of the Moon in this way. But a study in 2010 suggested that a naturally occurring nuclear explosion could have had enough power to hurl a Moon-sized chunk of the Earth out into space. The fact that it could happen doesn't mean it did – there is no evidence of such an explosion occurring.

Is it Earth's little sister?

As the Moon and Earth have a similar chemical composition, it's been suggested that both formed at the same time, near the start of the early solar system. Like Earth and other planets, the Moon would have formed from the collision of lumps of rock, dust and gas. The Moon's proximity to the much larger Earth could then have drawn it into orbit.

SIZE MATTERS

Our Moon is larger than the moons of nearly all of the other planets, relative to the size of the planet. Pluto is the only other body in the solar system with a moon of comparable relative size. This suggests the Moon might well have formed in a way different from the majority of moons in the solar system. Most are thought to be captured asteroids, seized by their planet's gravitational field after straying too close.

However, there are key differences which undermine this theory. In particular, the Moon has no iron core; it should have one if formed in this way. The particular geometric relationship of the Earth and the Moon is also not what we would expect if the theory were correct.

How to grow a planetary system

Since the work of Russian astrophysicist Viktor Safronov in the 1960s, scientists have believed that the planets all formed from a proto-planetary disc of dust and gas orbiting the Sun. Over time, clumps of matter collided and stuck together (because all matter exerts a

gravitational pull on other matter), forming larger and larger lumps. The heavier matter remained closer to the Sun and the less dense matter migrated outwards.

As larger chunks emerged they continued to grow, their greater gravitational pull attracting more and more matter, eventually forming planets and smaller planetisimals. Denser matter stayed closer to the Sun, so the inner planets are dense and rocky. Further out, planets are made of less dense materials – they are the gas and ice giants. After a lot of chunks had formed, but the planets had not yet cleared their orbits, there followed a period characterized by collisions between asteroids, planets and planetesimals.

Giant impact theory

What if, in the early days of the Earth, another new planet had crashed into ours with cataclysmic force? Could it have smashed to pieces and then coalesced into the Moon? This idea was proposed in 1975 by two American astrophysicists, William Hartmann and Donald Davis. They worked with calculations about the formation of solid bodies from the proto-planetary disc and came to the conclusion that there were likely to have been several smaller bodies alongside the Earth. Studies of the mechanics of the system suggested that whatever crashed into the Earth would have been between a third and half of its size.

TWO FOR THE PRICE OF ONE?

The same side of the Moon always faces Earth, so until we sent cameras and astronauts into space no human had ever seen the far side of the Moon. It turns out to be very different from the side we can see from the Earth. It has a thicker, more rugged surface with far more impact craters.

Computer models of the impact of Theia with the Earth show how two moons might have formed in the aftermath of the collision. They also suggest that after tens of millions of years, one moon would collide with the other at a speed of around 2–3km (1–2 miles) per second. This speed would not make a massive impact crater, but the second moon would be effectively smeared over the surface of the first. The thickness of the surface of the far side is consistent with this theory. So perhaps we once had two moons, but the one we're left with ate the other.

The birth and death of Theia

The hypothetical Mars-sized object that could have crashed into the Earth to form the Moon has been named Theia by scientists. (In Greek mythology, Theia was mother of Selene, the goddess of the Moon.) The original theory, as proposed in 1975, had Theia smashed apart as it struck the Earth a glancing blow, knocking out a large mass of material which mingled with the debris. The pieces of Earth and Theia formed a disc of debris orbiting the Earth which then clumped together to become the Moon.

The same, but different

Since astronauts gathered Moon rock in the 1960s and 1970s, scientists have been able to examine its chemical and geological composition in great detail. Before this, we could only look at meteorites from the Moon and they might have changed as they travelled through the Earth's atmosphere and hit the ground.

The evidence from Moon rock shows that Earth is richer in heavier metals, particularly iron, and water than the Moon. But the balance of oxygen isotopes (see box on page 188) is identical on the Moon and the Earth. The balance of isotopes generally varies between planets. That the Moon and Earth have the same proportions of oxygen-16 and oxygen-17 suggests a common origin. If the Moon contains parts of Theia, then so does Earth.

The close similarity suggests the Moon is largely made from bits of the Earth – or that Earth and the Moon are both made from the original Earth and Theia mixed together. Some of the differences between the Moon and the Earth are what we would expect. The Moon was created in a super-heated event that caused some components to evaporate or change from one form to another, so it is short of chemicals that

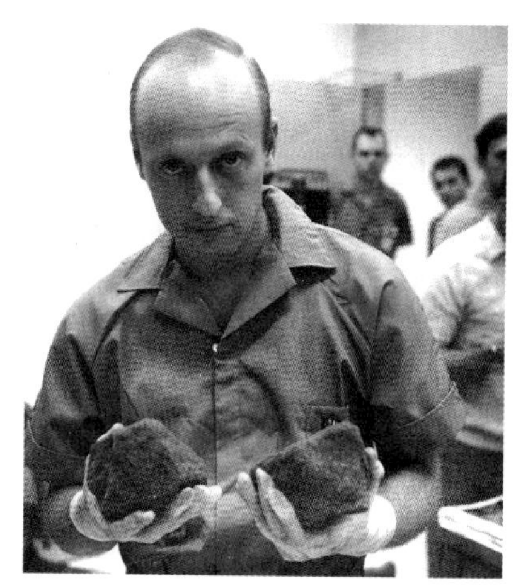

Bringing moon rock back to Earth has enabled us to analyze it fully.

ISOTOPES

The atomic mass of an element is the 'normal' number of nucleons (protons and neutrons) in the nucleus of an atom of that element. Some elements can exist in more than one form, though, with a different number of neutrons in the nucleus. These are called isotopes. An isotope is heavier than the normal element if it has any extra neutrons, and lighter if it has fewer than usual. Normal oxygen has 16 nucleons, of which eight are neutrons. The isotope oxygen-17 has 17 nucleons, of which nine are neutrons.

vaporize readily (including water). The lack of iron in the Moon could be because most iron had already moved towards the Earth's core by the time the collision took place and the chunk knocked out was crust and mantle and did not extend to the core.

Head-on collision?

When the giant impact theory was first suggested, it was thought unlikely that Theia could have collided head-on with Earth without entirely destroying our planet and creating a second asteroid belt in its orbit. More recently, better computer models have suggested that such a head-on collision could well have occurred without demolishing Earth. The most recent version of the theory supports this, holding that only a head-on collision would have produced enough energy to vaporize Theia and the Earth's upper layers, allowing their components to mix.

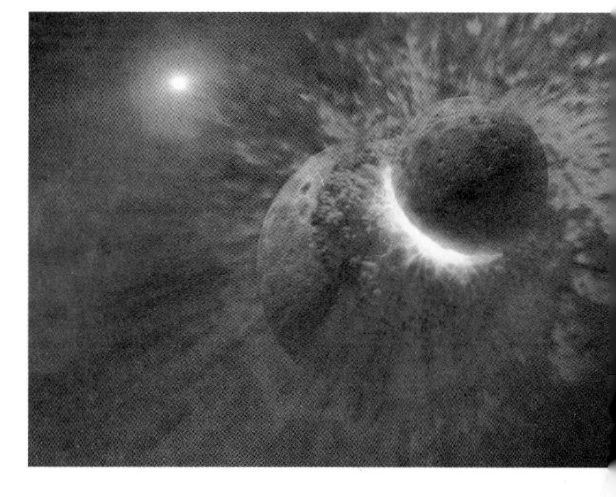

A new model of the giant impact, produced in 2016, suggests that Theia struck the Earth head-on with so much force that Theia and a portion of the Earth's crust and mantle vaporized. In the aftermath, the Moon formed from part of this mixture and some of it recondensed on the Earth. The near-identical isotopic composition of the two is explained by this theory better than by any of the other ideas suggested.

Asteroid scars

If the Moon was created by an impact event, material would have been scattered far and wide – it wouldn't only have clumped together to form the Moon. Some of it would have been thrown out towards the orbits of other planets and into the asteroid belt between Mars and Jupiter, where it would have collided with asteroids. Scientists looking at meteorites found on Earth which originally came from the asteroid belt have found some have chemical 'scars' dating from about 4.7 billion years ago. This could be evidence of the impact event, the age of the scars helping scientists to judge the age of the Moon. Current thinking puts the age of the Moon at around 4.5–4.55 billion years, with it forming only around 50–100 million years after the Earth had a solid surface.

Putting the Moon in its place

The 2016 version of the giant impact theory explains not only the composition of the Moon but also the position of its orbit, which

is not directly above the Earth's equator.

This posits a high-energy, head-on collision, that vaporized Theia and much of the Earth, which then mixed together so Earth and the Moon condensed from the same debris cloud. The Earth was left spinning very quickly, with a day only two hours long, and tilted on its axis so that it was pointing towards the Sun.

Slowly, the Moon slipped further away from Earth. When it reached the point where the gravitational effects between Earth and the Moon became less than the impact of the Sun, the Earth flipped nearly upright on its axis to its current orientation. The Moon was orbiting Earth at a high angle at this point and continued to move further away until it reached another transition point. Its orbit then dropped to its current angle of 5° above the Earth's equator.

This seems to cover all we need to have a Moon in its current position with its correct geological composition – and it gives both our planet and its Moon an exciting history.

Chapter 14
Recapitulation Theory

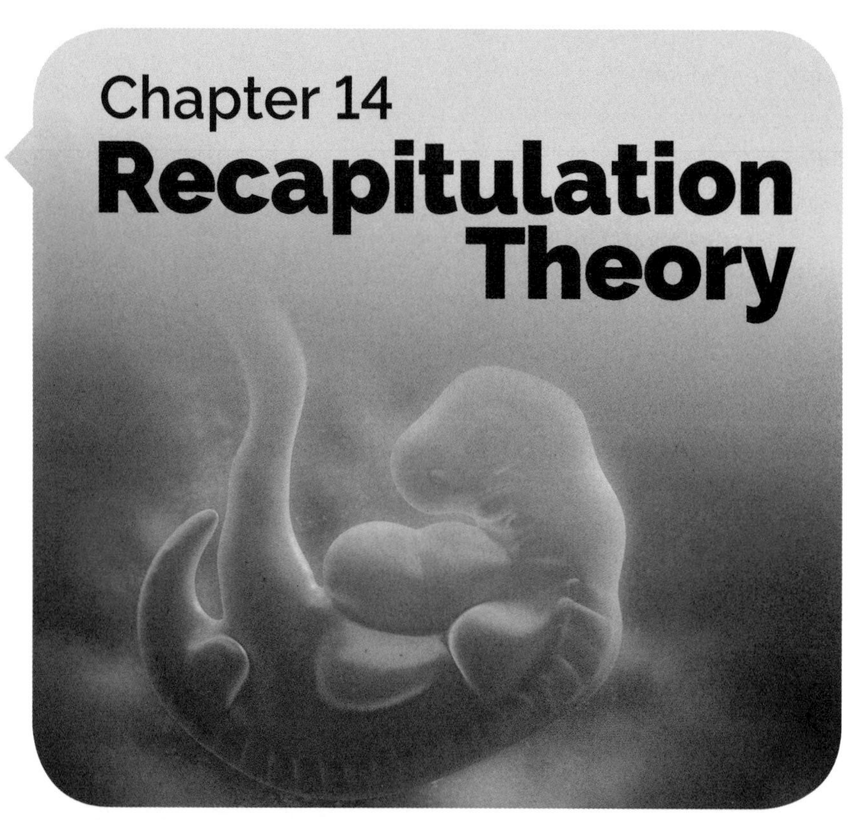

A developing embryo doesn't look much like the thing it is going to become – does it step through its evolutionary history as it grows?

Growing bodies and souls

People have been interested in how animals develop before birth or hatching for more than 2,000 years. In the 4th century BC, Aristotle suggested taking fertilized hens' eggs and opening one each day to see how the developing embryo gradually becomes a chick. The development of the embryo was not just interesting from a physiological point of view; it might throw some light on the tricky question of when the creature's soul arrives in its body.

THE IDEA

As an embryo develops in the womb or egg, it replays its evolutionary history, resembling its ancestors in a sequence that ends with its final form.

Parallelism

The German anatomist Johann Meckel is now all but forgotten, but in 1811 he had revolutionary insights into the development of the embryo. Before Darwin's theory of evolution (see page

52), Meckel was the first person to compare abnormal and normal development in the human embryo and to identify atavisms or genetic 'throwbacks' (see box). He proposed a theory of parallelism in which the development of the human embryo went through stages he felt were parallel to the structure of animals lower in the 'chain of being'. The chain, or *scala naturae* (ladder of nature), was an ancient idea that placed all life on Earth in a hierarchy, with plants at the bottom and humans at the top.

ATAVISMS

An atavism is the reappearance of a genetic feature from a previous generation of organisms. Examples include legs on snakes and short tails on humans. They appear when traits generally not expressed but still preserved in the DNA of an organism are, for some reason, expressed again in an individual.

Because animals were put into a hierarchy, Meckel could relate the stages of the embryo's development to a series of other animals without needing to suggest that humans developed from them. In the 1820s his ideas were extended to other animals and formulated into a theory by the French biologist Étienne Serres. The Meckel-Serres law proposed that embryos developed through stages parallel to fish, reptile, mammal and finally human forms. Each animal only passed through the stages 'lower' than its position in the chain, so

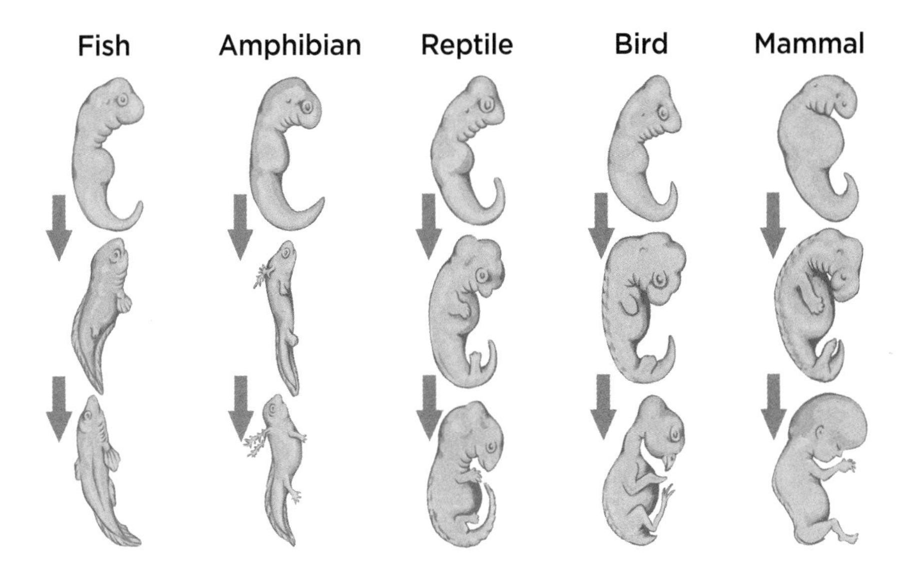

a reptile would look first like a fish and then stop at a reptile, but a mammal (other than a human) would look first like a fish, then like a reptile and finally a mammal.

Collecting organs as you go . . .

In the 1820s and early 1830s, Estonian embryologist Karl von Baer refined the theory further. He dropped the idea that the embryo goes through stages which recapitulate fully developed life-forms and instead had it accreting organs in a steady plod up through the hierarchy. So in its earliest stages the embryo resembles an *infusorium* (a very tiny aquatic organism); then it gains a liver and is like a mussel. When it develops a skeleton it has risen to the status of a fish, and so on, gradually making its way towards the human state. Again, this is a theory of parallelism rather than recapitulation as it doesn't suggest that humans have developed from these earlier life forms.

Eggs and embryos

Von Baer discovered that mammals develop from an egg cell, which grows inside the mother; he was the first person to see the human egg cell, though he did not give a description of it. (That the egg must be fertilized by a sperm to begin developing was not discovered until 1876.)

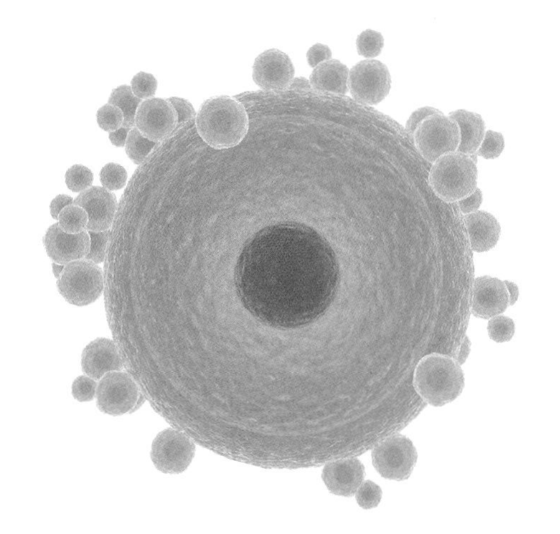

Von Baer developed four laws of embryology:

1 The embryo first develops the general characteristics of the group to which it belongs and later develops characteristics peculiar to its species.

2 The general structure of the embryo and relations between its parts develop before more specific structures appear.

3 An embryo does not grow to resemble other animals more closely as it develops, but grows more unlike them, separating itself.

4 The embryo of a higher animal form never resembles the adult of another animal form, even one lower in the hierarchy; it only ever resembles the embryo of another animal.

Since embryos develop increasingly specialized features as they grow, early embryos from different species look more alike than later embryos - they diversify as the specialized features of the organism develop. Consequently, following von Baer's parallelism, embryos develop in a sequence which parallels the classification of organisms into natural groups. So the first features to develop are characteristic of the phylum, then next of the class, then of the order, family, genus, and species in turn.

His studies of embryology led von Baer to believe in the transmutation of species - the idea that more 'advanced' animals had developed from less sophisticated ones over time. He even sketched an early phylogenetic tree showing how he thought animals were related to one another through their development, derived from his studies of their embryos. Even so, he rejected the natural selection element of Darwin's theory of evolution when it was published in 1859.

'Ontology recapitulates phylogeny'

The great German naturalist Ernst Haeckel proposed his biogenetic law in 1866. This was the first development of the idea since the publication of Darwin's theory of evolution, which Haeckel took on board. Instead of looking to the animal's place in a natural hierarchy of existing animals, Haeckel looked backwards to its genetic ancestry. He saw the development of the embryo replaying the evolution of the animal, with different stages of its development recalling adult forms in its evolutionary history. The theory of recapitulation is often stated in Haeckel's words: 'Ontogeny recapitulates phylogeny.' Ontogeny is the growth and development of an individual organism; phylogeny is the evolutionary development of a species.

Haeckel's ideas differed slightly from those of Darwin, who also looked to embryology for clues about

Darwin was interested in the clues embryology can give us about the evolution of species.

the evolutionary history of an animal. Darwin believed that the similarities in the early embryos of different species point to a common ancestor in evolution, and that the features which develop later in the embryo are those which have emerged most recently in the evolution of the animal. He was interested in comparing the embryos of different species and drawing inferences about the embryonic forms of previous (ancestor) species. Haeckel saw the adult forms of earlier ancestors recapitulated in the embryonic forms of current animals.

Haeckel (unlike Darwin) believed that evolution was directed and animals progressed in their development from less sophisticated to more sophisticated forms over time. He accepted Jean-Baptiste Lamarck's idea that animals change their form over time as a result of their behaviour and that these changes are passed on. So, for instance, giraffes eventually developed long necks because generation after generation stretched upwards to reach higher leaves and over time their necks grew longer and the increase was passed on. Darwin would explain the same trend by saying that giraffes which happened to be born with

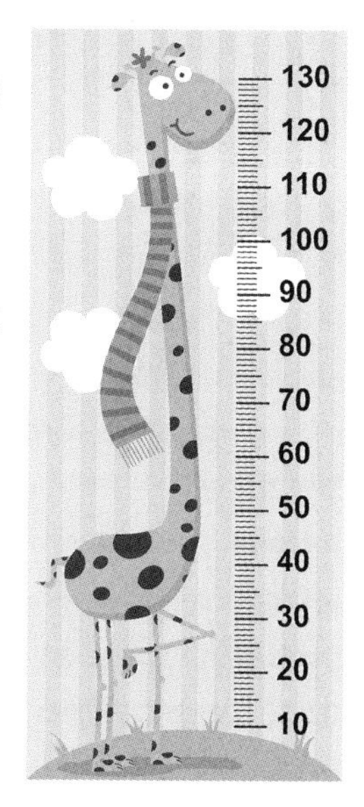

slightly longer necks were more successful at gaining food and more likely to survive and breed than short-necked giraffes. This means that long-necked giraffes are increasingly prevalent in the population of giraffes, eventually becoming the norm and a defining feature of the species.

Three assumptions

Haeckel's biogenetic law was based on three (unreasonable) assumptions:

1 The law of correspondence: in 'higher' animals, each stage of the embryo's development corresponds to a stage in evolution represented by an adult 'lower' animal. The human embryo appears to have gill slits at an early stage; Haeckel saw this as corresponding to fish having gills. (Mammalian embryos often have slits in the neck, called pharyngeal slits, but they are never part of a breathing structure. They develop as parts of the head and neck separate and divide, ready to form different structures, including the jaw and parts of the ear.)

2 Development proceeds by addition: evolution builds on 'lower' animals by adding more features and complexity so ontogeny proceeds by the same method. The embryo completes each stage in turn, copying a progression of simpler animals and then going on to add more advanced features.

3 The principle of truncation: as development proceeds by addition, pregnancy should get longer as an animal becomes more advanced. But it isn't as long as it logically should be. Haeckel assumed that in 'higher' animals the early stages of embryonic development are therefore speeded up.

There is clearly no stage at which the human embryo looks exactly like a fully grown 'lower' animal. Haeckel explained this away by saying that truncation produced such inconsistencies. It was a bit of a botch, and his ideas soon attracted critics.

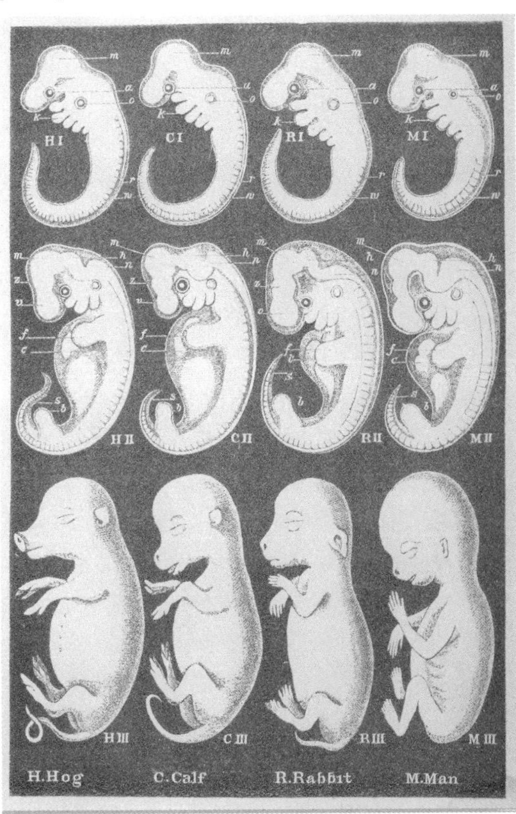

Haeckel's drawings of embryos attempt to show recapitulation.

In 1874, Haeckel supported his theory with drawings of embryos of fish, salamanders, tortoises, chicks, pigs, cows, rabbits and humans at various stages of development, in which he made the humans look as much like the other animals as possible. It was still not entirely persuasive and many other embryologists found far less similarity than Haeckel claimed to see.

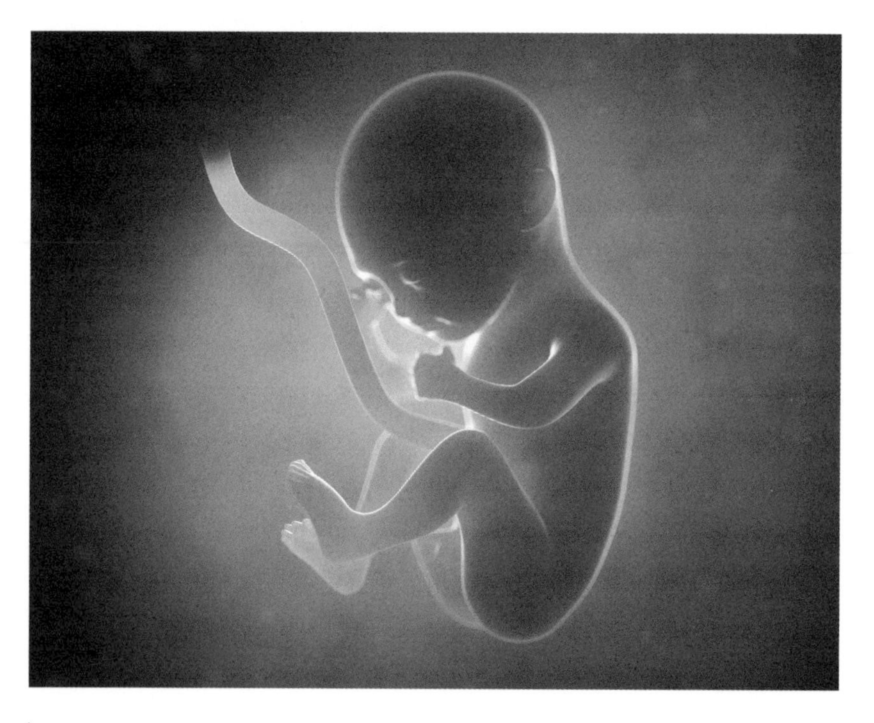

Haeckel's theory has now been entirely discredited. Darwin's view of early similarity between embryos of different types of animal has prevailed. There is some similarity in the early stages of development of animals that share common ancestors, but there is no process of recapitulation. Many animals share common basic structures and body plans, so this naturally leads to some similarities as they develop, but the embryo differentiates increasingly as it develops features particular to its type and its finer details of final shape emerge.

Von Baer's and Darwin's approaches both compared the embryos of different types of animals rather than the embryo of one with the fully grown form of others. This was the approach that continued into the 20th century.

OUTSIDE THE EMBRYO

The general principle of Haeckel's theory has been applied in completely separate disciplines, including psychology, education and even art history. For example, in child development it is used to suggest that a child acquires types of knowledge in the same sequence that humankind as a whole acquired knowledge. This was first proposed by Herbert Spencer in 1861 – five years before Haeckel publicized his theory. Spencer saw education as recreating civilization 'in little' and stepping through it in order to the present state of knowledge and 'civilization'.

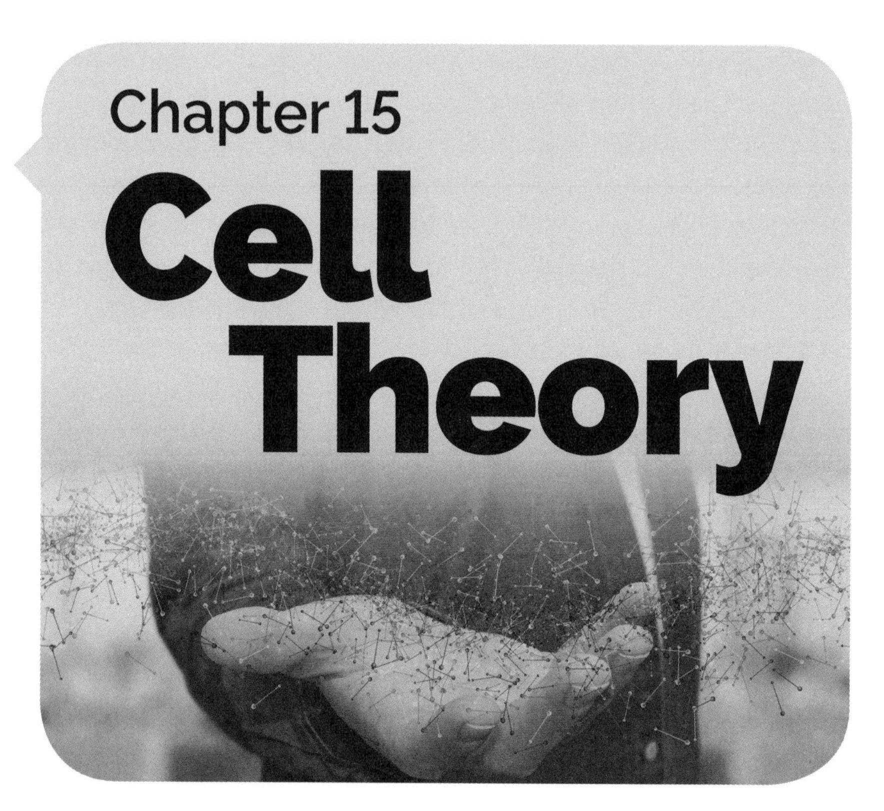

Big things are composed of smaller things. But the building blocks of animal and plant bodies, including our own, are complex and surprising.

Eye-opening development

Before the development of the microscope in 1600, no one knew that bodies are made of structurally and functionally similar building blocks. Even once the microscope had revealed the structures, it was another 200 years before the implications were clear.

THE IDEA

All living organisms (except viruses) comprise one or more cells. These are self-contained units, and form the basic unit of life. Cells reproduce, so all cells come from pre-existing cells.

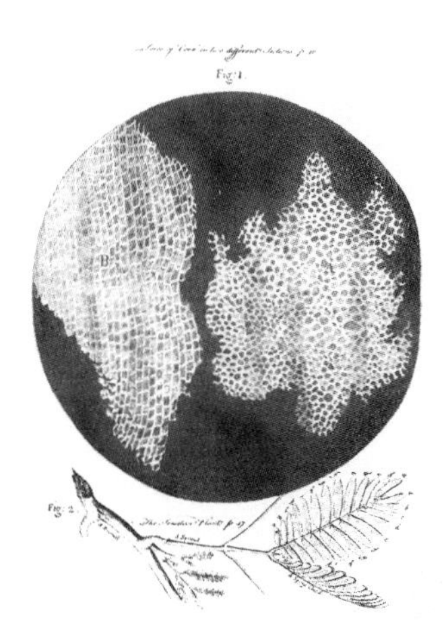

Cells and monks

In 1661, King Charles II of England commissioned a book on insects seen through the microscope. The man he commissioned was busy – he was already a professor of astronomy – and he soon passed the task to a keen young scientist and draughtsman. That young man was Robert Hooke, and he

Cells as drawn by Hooke in Micrographia.

became one of the greatest scientists of the age. The man who gave up the task was Christopher Wren, who later went on to design the new St Paul's Cathedral after the Great Fire of London.

Luckily for biology, Hooke was so enthusiastic that he went beyond the commission to look at all kinds of other things under the microscope. The book he produced of his drawings in 1665, *Micrographia*, is one of the greatest science books ever published. One of the less dramatic, but most important, drawings in it is of the microstructure of a piece of bark from the cork tree. In it the tiny building blocks of a living organism were clear to see for the first time.

Hooke saw solid walls surrounding what seemed to be empty spaces. He referred to these as pores or 'cells' since they recalled the small rooms called cells that monks occupy in a monastery. He calculated that there are 1,259,712,000 cells in a cubic inch of cork (equivalent to 76,872,000 in a cubic centimetre).

The name stuck – we still call them cells. But we no longer think of them as empty spaces in a solid structure. We now

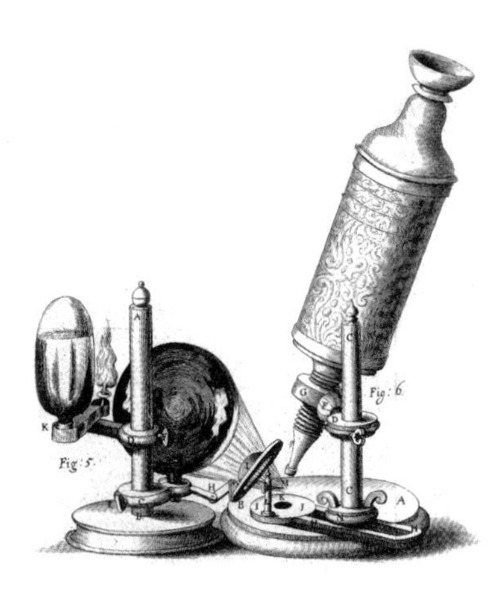

Hooke's microscope.

know that cells are self-contained units, pressed closely against one another. The solid fretwork that Hooke saw is the walls of the cells, and the important business goes on in the 'spaces'. Indeed, the whole of life goes on in the spaces.

Animalcules

Micrographia inspired many people, including a Dutchman called Antony van Leeuwenhoek. He began grinding his own lenses, which were much better than those he could buy, and began to look at all kinds of things from the world around him. They included pond water, body fluids and the gunk scraped from his teeth. In his samples he saw what he recognized as tiny free-floating organisms, which he called 'animalcules'. He considered movement to be a quality of life, so concluded that they were alive. He was the first person to describe living unicellular organisms, in 1674.

> ***'I then most always saw, with great wonder, that in the said matter there were many very little living animalcules, very prettily a-moving. The biggest sort . . . had a very strong and swift motion, and shot through the water (or spittle) like a pike does through the water. The second sort . . . oft-times spun round like a top . . . and these were far more in number.'***
>
> Antony van Leeuwenhoek, on the gunk from his teeth, 1683

It would be nearly 200 years before anyone saw a connection between the monks' cells that Hooke had drawn and Leeuwenhoek's animalcules.

Plants and animals

It was easier for the early microscopists to see cells in plants than in animals, as plant tissue is firmer and more robust. Animal tissue tends to squash when sliced and deteriorates more quickly. But no one was looking to compare them: plants, animals and the tiny creatures revealed by the microscope were considered to be entirely separate realms of life, occupying different roles in Creation. Why would they be similar?

Eventually, better procedures for preparing samples were developed and animal tissues began to reveal their secrets. In 1824, the French scientist Henri Milne-Edwards suggested that all animal tissue is made up of 'globules' of uniform size. The principle was correct, but the detail – that cells are all the same size – was very wrong.

Spaces or structures?

In 1804, the Swedish naturalist Karl Rudolphi recognized that each cell has an independent wall. This was significant insofar as it established that cells are separate units packed closely together rather than a composite structure like honeycomb or coral. The French biologist Henri

Dutrochet took this the necessary step further, proposing for the first time that plant and animal cells are comparable – indeed, fundamentally equivalent – as the building blocks of living bodies.

'It is clear that it constitutes the basic unit of the organized state; indeed, everything is ultimately derived from the cell.'
Henri Dutrochet, 1824

Putting the pieces together

Cell doctrine as it is now known did not come from Dutrochet's statement but from an after-dinner conversation between two German biologists. They were the botanist Matthias Schleiden and the physiologist Theodor Schwann. Both had been independently observing cells in their subjects of study and as they talked they realized they had been seeing the same things: cells as structural units of organization in plant and animal bodies. The logical conclusion was that all living things are made up entirely of cells. Schwann published this, the first part of cell theory, in 1839. He made no

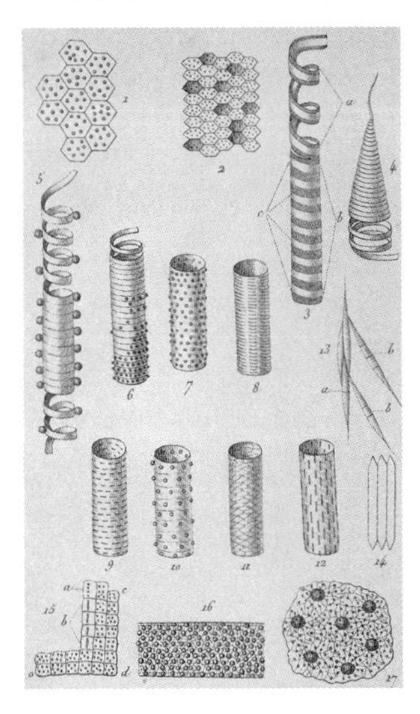

From Dutrochet's Recherches Anatomiques, *1824.*

reference to Schleiden's contribution, even though Schleiden had already published, the previous year, his finding that plant bodies are made up of cells. And in 1837, too, the Czech physiologist Jan Purkyně had already said that animal bodies are composed mainly of cellular tissue which 'is again clearly analogous to that of plants'. The idea was in the ether; Schwann brought it together in a clear formulation and is credited by history with the discovery that plant and animal bodies alike are built from a basic functional unit – the cell.

> ***'A common principle underlies the development of all the individual elementary sub-units of all organisms.'***
> Theodor Schwann, 1839

The first two tenets of cell theory were set out by Schwann:

- All living organisms are composed of at least one cell.
- The cell is the fundamental unit of life.

In fact, not quite all tissues

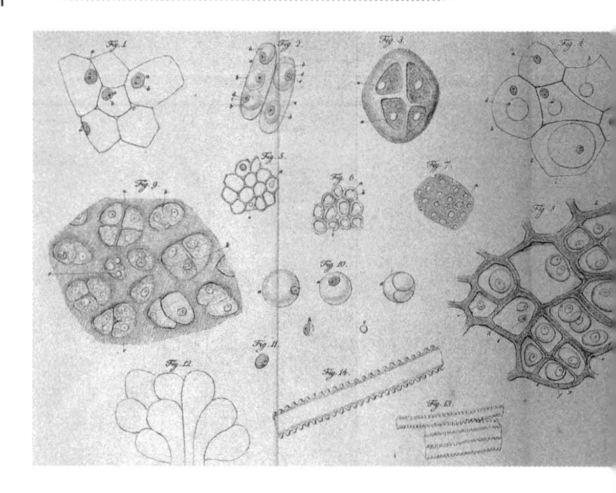

From Schwann's Mikroskopische Untersuchungern.

were recognized as cellular. Nerve cells (neurons) were not identified as cells until later, when microscopy techniques had improved. Neurons come in very many shapes and sizes and do not share the usual form of animal cells. They are also particularly difficult to stain and distinguish.

Cells from cells

If cells are the basic building blocks of life, as cell theory proposes, there has to be a mechanism for making more cells since clearly, as a plant or animal grows, it has to add more cells to its body.

Schleiden believed the cells just crystallized into existence, either within other cells or outside them. The theory of spontaneous generation – that some organisms can just spring into life from inanimate material (see page 26) – had taken a few knocks at the time, but it was still widely believed that very small things could appear spontaneously. There was no compelling reason to reject this theory.

Even so, Belgian botanist Barthélemy Dumortier had described what he called 'binary fission' in plant cells in 1832. He had seen a partition grow between an old cell and a newly forming cell and recognized this as 'a perfectly clear explanation of the origin and development of cells'.

Dumortier is not generally credited with discovering how cells

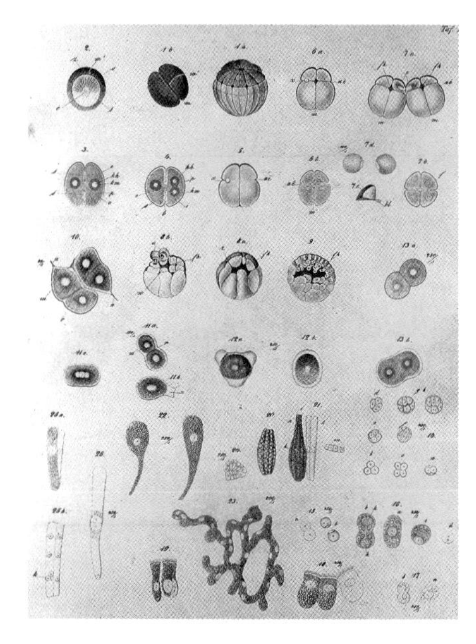

From Remak's work on cell theory.

are formed. That distinction goes to Rudolf Virchow. But Virchow took the information from another biologist, Robert Remak, who in 1852 had published his findings that animal cells reproduced by binary fission – by splitting in two. Remak worked on embryos, which are veritable cell factories, having to produce enough cells for an entire organism starting from a single fertilized egg cell.

Virchow publicized this finding without mentioning that Remak had already discovered it and published it. In one of those unfair accidents of history, it is Virchow's name that is most often associated with the third tenet of cell theory:

- All cells come from other cells.

More bits of theory

As more detail about cells and their behaviour became available, theories proliferated about their chemistry and how they work. But the three basic principles remained unchanged.

There is now much more to cell theory than the original three tenets. Modern cell theory adds:

CATCHY PHRASE CAUGHT

Virchow is usually also credited with the catchy slogan '*Omnis cellula e cellula*' (all cells from cells), but this too pre-dates his work; François Raspail first used the phrase in the 1830s.

- The activity of the whole organism depends on the total activity of all its independent cells.
- Energy flow occurs within cells, the result of the cell's metabolism and biochemical processes.
- Cells contain genetic material: DNA in the chromosome(s) and RNA in the cell nucleus and cytoplasm.
- In organisms of similar species, all cells have similar chemical composition.

Genetic material is copied and transferred to new cells as they form, so passing on the 'instructions' for making and running the organism from old cells to new. The flow of energy between cells is handled by proteins which can control the movement of chemicals across the cell wall.

Although all cells contain similar chemicals, they occur in different proportions and are arranged differently. This means an organism can have very different types of cells. Plants typically have 10–20 different types of cells, but animals can have 150–200 different types.

Two types

There are two broad categories of cells: prokaryotic and eukaryotic. Prokaryotic cells are the simpler type and are found in prokaryotic organisms such as bacteria. They have no nucleus or organelles (self-contained cell structures) and their genetic material is stored as a chromosome in the protoplasm. Eukaryotic cells are found in eukaryotes – more complex organisms, including all plants and animals. These cells have a more sophisticated structure, with (usually) a nucleus and various cell organelles, which include mitochondria (the energy 'power house' of the cell) and the chloroplasts that are responsible for photosynthesis in green plants.

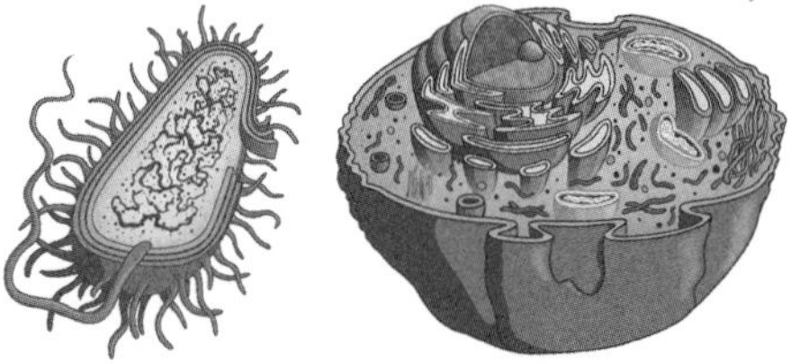

Prokaryotic cell (left) and eukaryotic cell (right).

Cells inside cells

The similarity between these organelles and bacteria was noticed in the 19th century, but there was no way of investigating it. In the 1960s, American evolutionary theorist Lynn Margulis suggested that mitochondria were originally separate single-cell organisms that lived alongside other cells in a symbiotic relationship (meaning they benefited one another). Eventually the mitochondria become absorbed into plant and animal cells. The same is likely to be true of chloroplasts and other organelles.

Advances in knowledge and technology in the 1970s made it possible to investigate the genetic material (DNA) inside cells. Mitochondria have their own DNA loop, separate from the chromosomes in the nucleus of a cell. This DNA is now thought to be similar to that of a bacterium that causes typhus. The DNA in chloroplasts is similar to that of cyanobacteria, a simple type of algae which was the first organism to photosynthesize. It seems that every eukaryotic cell is a fusion of several other types of cell.

HOW MUCH OF YOU IS ACTUALLY YOU?

It's commonly said there are ten times as many microbial cells in the human body as human cells, but in 2016 this was overturned – you're only about 50 per cent microbial cells. Even that's a bit creepy. An 'average' person (in scientific terms, a man aged between 20 and 30 years, weighing 70kg (155lb) and 1.7m (5ft 7in) tall) has about 30 trillion (30,000,000,000,000) human body cells and about 39 trillion bacteria. There will be wide variations between people, with some being more than half bacteria and some more than half human. Add to that all the mitochondria fully incorporated into the human cells and we are all veritable colonies of other beings!

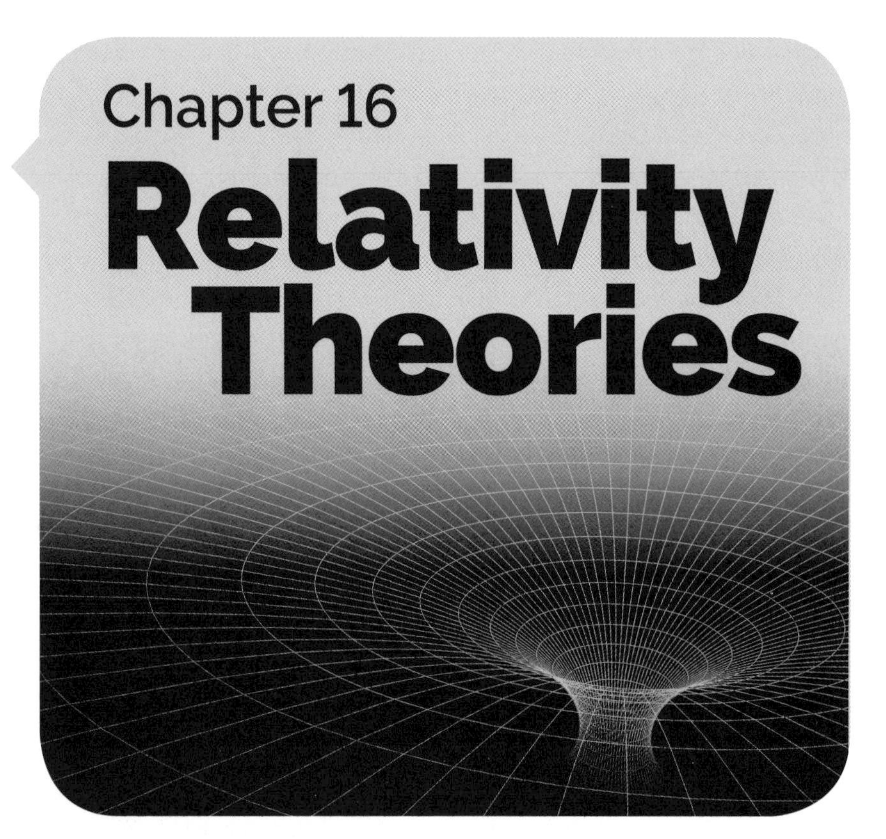

Chapter 16
Relativity Theories

Einstein's two theories of relativity, special and general, are among the most widely recognized scientific theories – and the least understood.

So good he did it twice

Einstein published his first relativity theory in 1905. Although the special relativity theory was mind-blowing and game-changing in itself, Einstein was keenly aware of its limitations. He set to work immediately on improving and extending it until, ten years later, the general relativity theory was ready.

THE IDEA

Einstein produced two relativity theories. The first only applied in special circumstances – to states in which the effects of gravity are negligible.

1

The special theory of relativity describes space-time: the fabric of space and time are inseparable, and consequently both time and distance are relative.

2

The general theory of relativity provides a new model for gravity, which is described as curvature of the space-time continuum.

It's special

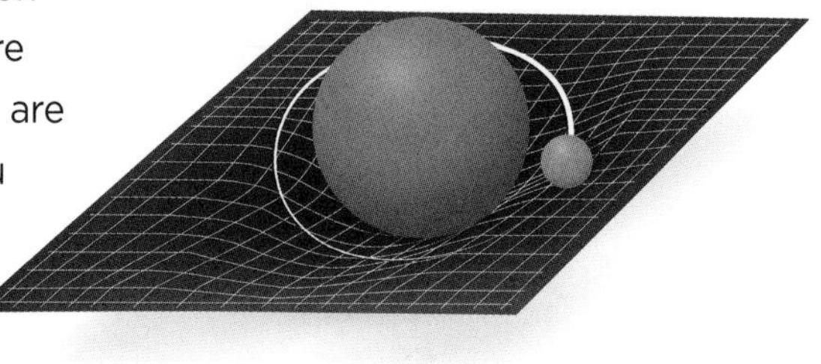

Whether or not you are considered tall depends on who or what you compare yourself with. Even if you are very tall for a human, you are not tall relative to a giraffe. However, if your hair is black, that is absolute – either it is black or it is not. We are used to the notion that some qualities or 'facts' are relative and others are absolute.

Einstein became interested in the relativity of things in terms of time and space. If you look out of the window at two objects, one on the left and one on the right, it's easy to see that you could walk to a position on the other side of the objects where they would have switched places in relation to you– the one on the right would be on the left, and vice versa. So position in space is relative to the position of the observer.

Where the observer is and how they are looking is called the 'frame of reference'. If you stand in your house and look out of the window at two cars moving in the same direction, your frame of reference is not the same as if you were in one of the two cars. To you, it might appear that one car is travelling at 50km/h (31mph) and the other at 70km/h (44mph). To a passenger in the first car, the second car will appear to be going at 20km/h (12mph). But none of you, in the house or in a car, is aware of the Earth moving through space.

This works with positions as well as speeds. If you are inside the car and you throw a sweet to another passenger, the sweet appears to move just across the car. But it has also moved some distance along the road on which the car is travelling – plus the

MOVING IN SPACE

The Earth is constantly spinning on its axis. At the Equator, the speed of movement is about 460m (1,500ft) per second. In addition, the Earth is orbiting the Sun, covering the whole distance in a year, moving at an average speed of 30km (18 miles) a second. On top of that, the solar system is whizzing through space, orbiting the centre of the Milky Way at 220km (136 miles) a second. And then the Milky Way itself is moving at around 1,000km (620 miles) a second towards an area of space called the Great Attractor. In all cases the Earth is moving, but 'relative to what?'

The satellite COBE (Cosmic Background Explorer) has made measurements which show the Earth is moving towards the constellation Leo and away from the original source of the cosmic background radiation (see page 20) at a rate of 390km (240 miles) per second. Relative to the origins of the expanding universe is as absolute as we can get.

distance covered by the Earth in that time. While you are observing something in your own frame of reference, it's possible to compare positions and speeds with some confidence that you will get a meaningful result. But the result is only meaningful for an observer sharing your frame of reference – in this case, an observer who is also in the car, if you are throwing the sweet. To an observer on Mars, the distance the sweet has covered within the car is an infinitesimal proportion of the distance it has really covered, which includes the movement of the car, the revolution of the Earth and the orbit of the Earth around the Sun.

EINSTEIN STARTS THINKING

Few theories spring from nowhere at all; they generally follow an attempt to solve a problem of some kind. For Einstein, the problem which set him along the road to special relativity was that the laws of electric and magnetic forces described by James Clerk Maxwell's equations in 1861–2 did not work with the prevailing notions of space and time. The laws seemed quite clearly right, so he assumed that it was the ideas of space and time that were wrong and set out to improve them.

Maxwell demonstrated that electromagnetic radiation operates as waves, and that their speed is the same as the speed of light, so it's reasonable to assume that light is a form of electromagnetic wave. Special relativity fits perfectly with Maxwell's laws.

Who's moving?

Einstein began by thinking about observers in an inertial frame of reference. An example would be astronauts on two spaceships unaffected by the gravity of any stars, floating freely at a constant speed – they have no acceleration. The theory of special relativity relates to observers like this. It addresses which things can be said to be relative and which absolute to these observers.

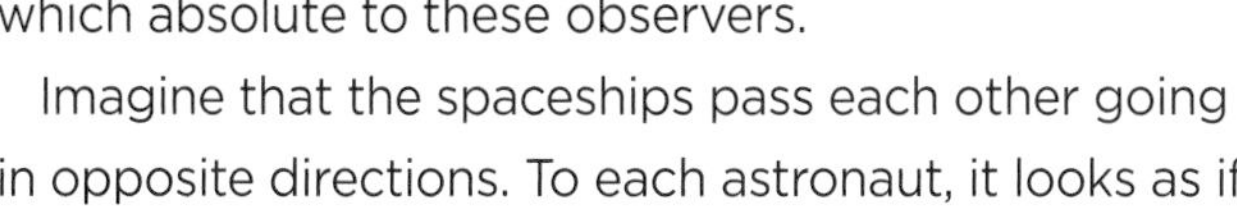

Imagine that the spaceships pass each other going in opposite directions. To each astronaut, it looks as if the other spaceship is moving and their own is still. Without making reference to some external landmark, it's impossible to say which is moving. Even then, the landmark might be moving – how would we know? You might well have experienced something similar yourself. If you sit in a stationary train next to another stationary train and one of them starts moving, it's hard to tell by just looking at the other train which of them is in motion. We have other cues – sound, the feeling of vibration, whether other objects seen through the window seem to move or not – which help us to work it out. But without these to refer to, it would be impossible to tell.

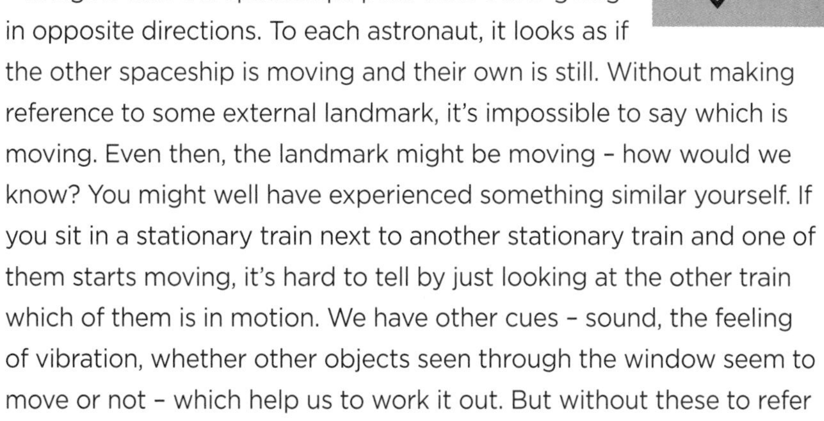

Movement and speed, then, are relative. This suggests that distance and time might be relative and it turns out to be true. The best we can say is that for any two observers in the same frame of reference, time and distance will appear to be the same.

Both at once? When's 'once'?

Suppose that the astronaut on spaceship A witnesses two events which seem to be simultaneous – they appear to happen at exactly the same time. The astronaut on spaceship B does not see them as simultaneous because her spaceship is in a different place, closer to one event than another. The light from each event takes a different length of time to reach each spaceship. There will always be a point in space at which the two events appear simultaneous, but at other points they won't.

We could say they are only 'really' simultaneous if the point at which they appear to be so is exactly halfway between the two. This is complicated enough if the two observers are stationary. If they

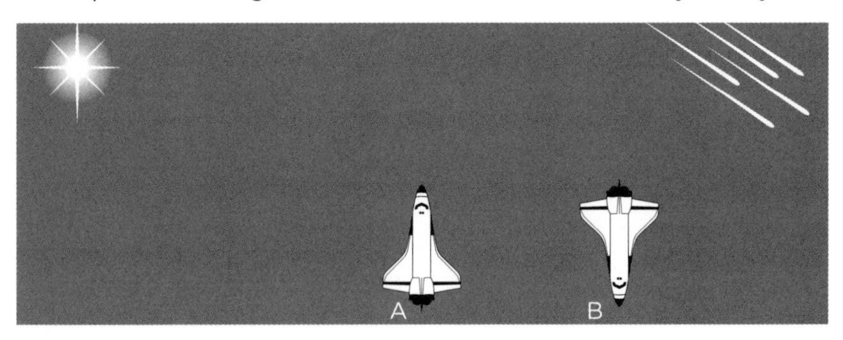

are also moving relative to each other and the observed events, it becomes even harder to determine whether (or when) events are simultaneous. Time and position become inextricable. Einstein stopped talking about space and time and separate dimensions and spoke instead of space-time as a single, four-dimensional continuum.

As fast as light

The one thing that is absolute and unchanging regardless of frame of reference is the speed of light. At 299,792km/second (186,282 miles/second) it sets the speed limit for the entire universe - nothing can go faster. The reason nothing can go faster than the speed of light is given by Einstein's famous equation:

$$\mathbf{E = mc^2}$$

where E is energy, m is mass and c is the speed of light.

We'll see why later.

Shrinking and stretching time and space

As we approach the speed of light, both distance and time become more obviously relative. Essentially, time slows down at very high speeds and distances contract. These effects are called speed dilation and length contraction.

Suppose an astronaut in one of our hypothetical spaceships directs a laser at a mirror on the ceiling of the craft, which reflects the beam

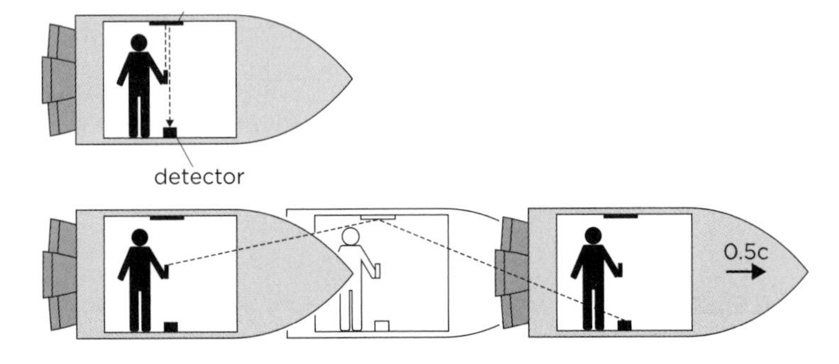

back to a detector on the floor. From inside the craft, it looks as though the light has travelled a short distance, from the laser to the ceiling and back down to the floor.

This won't take very long at all – let's call the time it takes 't', where 't' is a very tiny fraction of a second.

As this spaceship speeds past the other, our second astronaut looks out of the window and sees the experiment. To her, it looks as if the light has gone much further in the same time, as the spaceship is moving as well. The light seems to follow a slanting path up to the ceiling and back down. But the length of time the light of the laser actually takes to follow its path must be a fixed value. The only way the speed of light can stay the same is if time slows down at this high speed, allowing the light to follow its path at the right speed. In fact, if we do some complicated maths, it's possible to show that the speed of light is the same for both observers.

Einstein Tower in Potsdam, Germany, is an observatory completed in 1921 specifically to test the theory of relativity.

Energy and mass

It's not only distance and time that get jumbled up in the special relativity theory. Mass and energy turn out to be the same thing, as Einstein's equation suggests.

As an object travels very fast, approaching the speed of light, it gets heavier and heavier. This means it takes more and more energy to accelerate it a bit more. Adding energy automatically increases the mass, so if we try to accelerate a body, it gains energy (kinetic, or movement, energy), so its mass increases. It would become infinitely heavy at the speed of light, so would take an infinite amount of energy to accelerate to that point – therefore it's not allowed, according to the laws of physics.

There is, though, something rather useful about the equivalence of mass and energy: one can be converted into the other. Even saying 'converted' is misleading as they are the same thing, just differently experienced. All matter is basically just energy in little packets (quanta) following the rules for matter. But with enough effort it's possible to get useful energy from matter, by splitting up (or sticking together) atoms. Nuclear fission releases energy by breaking atoms apart. Nuclear fusion releases energy by forcing atoms to fuse together. It powers the Sun and other stars.

Not enough

Although he had united mass and energy, undermined our concept of time and distance and fixed the speed of light as an absolute value no matter where it's viewed from, Einstein still wasn't entirely satisfied. His theory only worked in the special circumstances set out at the start – the inertial frame of reference, which meant that the objects involved are

APPLES STILL FALL DOWN

The laws that Newton formulated still work in almost all situations. His theory is still good enough for launching spaceships and any other practical applications we can think of. But it falls down in extreme situations – it is incomplete rather than wrong. Einstein's new version, presenting gravity not as a force but a kind of geometry, works in a wider range of situations.

not accelerating. Special relativity sorted out some problems in the physics of the day, reconciling the laws governing the movement of bodies described by Galileo and Newton with the behaviour of electromagnetic radiation described by Maxwell. But there was a problem – it didn't take account of gravity. As we have seen, gravity is equivalent to acceleration (see page 122) and Einstein's theory only worked in situations where there is no acceleration.

From 1907, Einstein turned his mind to extending his theory from the special situation of the inertial frame of reference to encompass gravity (accelerating bodies). It meant throwing away much of classical physics and, in particular, Newton's theory of gravity.

An object with a lot of mass distorts space-time around it.

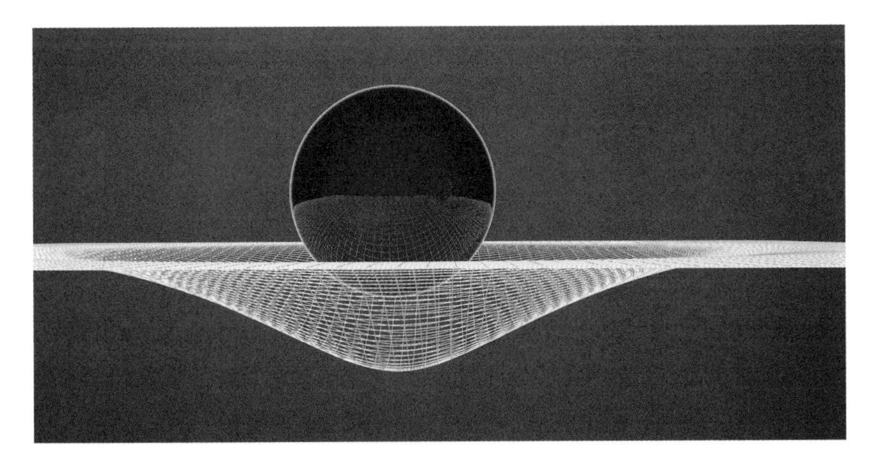

Rewriting gravity

Einstein drew on new maths developed in the 19th century to come up with a description of gravity as an effect of the fabric of space-time. Instead of a force operating between bodies with mass (see page 114), he described it as curvature in space-time which affects the way bodies move in relation to one another.

A common analogy is the surface of a trampoline. Imagine a heavy person standing on a trampoline, causing a dent. If you dropped a ball on to the trampoline, it would roll towards the person, following the slope of the surface. In Einstein's model, space-time acts rather like the trampoline, but in four dimensions. We can just about imagine it in three dimensions, but four dimensions defies visualization!

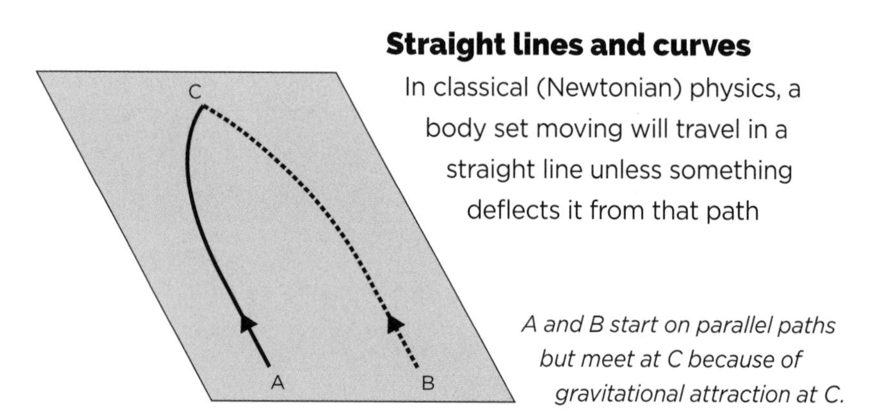

Straight lines and curves

In classical (Newtonian) physics, a body set moving will travel in a straight line unless something deflects it from that path

A and B start on parallel paths but meet at C because of gravitational attraction at C.

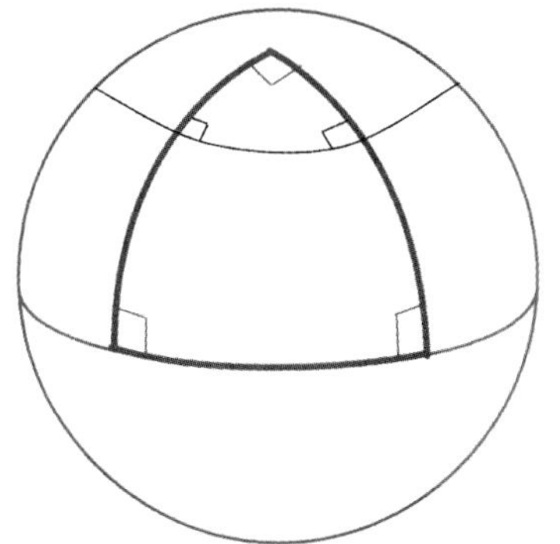

('something' can include gravity). We define a straight line as the shortest line that can be drawn between two points. Two bodies setting off on parallel straight paths will never meet unless something deflects one or both of them.

There is only one straight line that can join two points in a flat plane. On curved surfaces, though, the shortest line between two points is a geodesic, a line which follows the curve. There can be many equivalent geodesics. For instance, an infinite number of geodesics join the North and South Poles and all are the shortest path over the surface. The geometry of curved surfaces is very different from that of flat surfaces. For example, the angles of a triangle drawn on a globe often add up to more than 180 degrees.

While Newton's version of gravity forces bodies to leave their straight paths when they are attracted by another body, Einstein's has them continue on a straight (shortest) path but alters the shape of space-time to manipulate them into a different position. In the presence of a heavy body, a light one will accelerate towards it or go into orbit around it, but still follow the straightest possible path.

Gravity does not exert a force in the Einsteinian universe, it simply changes the way bodies move. And as bodies move, they in turn affect the curvature of space-time, and that further affects the movement of bodies, and so on.

Proving it

The first fundamental proof of the theory of general relativity was in accounting for oddities in the orbit of the planet Mercury. The laws derived by Newton and Kepler would have Mercury move in a regular ellipse with the Sun at one focus, though the presence of the other planets would disturb this a little.

Einstein's relativity equations give Mercury a shifting orbit, with it moving slightly on each circuit. Again, the other planets have to be taken into account.

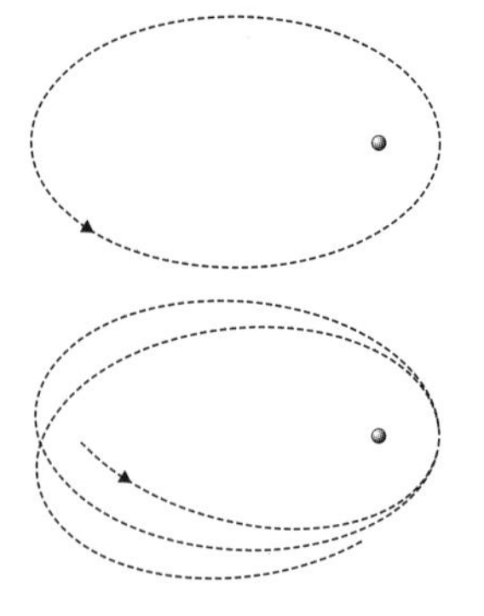

Detailed measurements of the orbit of Mercury show that it is far closer to the predictions of Einstein's model than to those of the classical model.

The most important proof came in 1919 when observations of a solar eclipse demonstrated that gravity can bend light, as Einstein predicted.

The idea of Mercury's regular ellipse was challenged by Einstein's revelations.

In a 1919 solar eclipse, it was demonstrated that light is bent by the Sun's gravity.

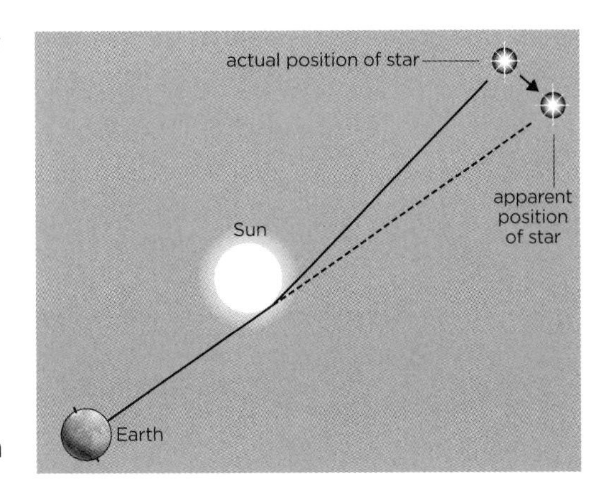

British astronomers photographed the position of a star close to the Sun during the eclipse and compared it with photographs of the same star at night (when the Sun is elsewhere in relation to it). The position of the star relative to more distant background stars showed that light from it is deflected by the gravity of the Sun. This effect is gravitational lensing: light from a distant object alters its path around a massive object and so it appears to be in a slightly different position from its actual position. The massive object (such as the Sun) acts as a lens.

Black holes and wormholes

The general relativity equations predict some bizarre possibilities. Two of the more intriguing are black holes and wormholes.

Black holes are not really holes – in fact, they are rather the opposite of holes. Far from being areas with nothing in them, they are areas with a very great deal of matter, immensely compressed. As a black hole has a very large mass occupying a very small volume, it's very dense. Its gravity is so strong – that is, it deforms space-time in so extreme a manner – that anything which approaches too closely inevitably accelerates into it and is absorbed. A black hole may form when a star collapses in on itself at the end of its life and there are

Wormholes are hypothetical distortions of space-time that might provide tunnels or shortcuts.

also thought to be super-massive black holes at the centre of many or all galaxies.

Contrary to popular belief, a black hole neither sucks in everything in its vicinity nor wanders through space sucking in other stars and planets. The gravitational effect of a black hole is determined by its mass, and as a black hole produced by a collapsing star has the same mass as the star had originally, its planets will not be automatically sucked in as they were not sucked into the star previously.

Black holes have not been observed directly because, by their very nature, they don't emit light or other electromagnetic radiation that we can detect. Their presence can be deduced from other evidence, though. Wormholes, on the other hand, might or might not exist. As proposed, they are like a black hole with two ends which provide a short-cut through folded space-time. They are unlikely to be useful for interstellar travel as anything entering a wormhole would probably be destroyed.

A unified field theory

Einstein was not entirely satisfied with his general relativity theory either. He wanted to be able to reduce everything in the universe to fields and to unite general relativity with electromagnetism. He spent the remainder of his life, from 1917 to 1955, trying without success to solve the problem.

Chapter 17

Gene Theory

How does your body know how to make itself? And what determines your resemblance to your parents and siblings?

Your mother doesn't matter

You would think the family similarities might have been a clue to the earliest thinkers that somehow we take characteristics from both of our parents. But the Ancient Greek philosopher Aristotle thought all the information about inherited features came from the father and the mother provided only the physical material needed for the child to grow. Aristotle's theory would naturally produce a race of boys identical to their fathers – and this clearly hasn't happened. Aristotle said that differences, including being a girl or looking like the mother in any way, were the result of faulty growing. So the mother's contribution to the growing offspring was to make it go wrong. That remained the dominant view for around 2,000 years.

THE IDEA

The blueprint or recipe for a living organism is coded in genes, which are chunks of a complex chemical, DNA. They are copied and passed on when organisms reproduce, so organisms share features with their parents.

Both together

Only in the 19th century did it become apparent that both parents make a contribution. Sea urchins were the clue. First, Alphonse

Derbès in 1847 and then Oscar Hertwig in 1875 discovered that both a sperm and an egg are needed and that the sperm must fertilize the egg before the offspring can grow. It still wasn't clear what each brought to the party, but at least they were now both invited.

On two tracks

The 19th century was an active time for biologists. Darwin's theory of evolution, published in 1859, inevitably prompted people to think about how the characteristics he talked about were passed on and changed. It would take a while, and for two different paths to meet, before the puzzle could be solved. On the one hand, the patterns of inheritance – how features pass from one generation to another, and which ones are passed on – emerged from statistical studies. On the other hand, the chemical mechanisms within the cell for storing and passing on characteristics emerged slowly through microscopic and biochemical research. It wasn't clear how they were going to intersect until the 1920s.

The monk in the pea garden

At the same time that Darwin was working on evolution in England, a Franciscan monk in Moravia (now the Czech Republic) was growing peas. Over eight years, 1856–63, Gregor Mendel grew 29,000 pea plants and studied the patterns of inheritance of characteristics, including flower colour and pod shape.

Mendel found that some characteristics appear in around a quarter of the offspring even if neither parent plant had that characteristic. By turning to statistics he found a solution. He suggested invisible 'factors' (now called genes) carry information between generations and that for each inherited trait there are two 'forms' (now known as alleles). The alleles are passed on in pairs, one of which is dominant – it trumps the other if there is one of each. The other is recessive.

Using the example of flower colour, there is an allele for white flowers and one for purple flowers. Purple is the dominant allele, so if a plant has either one or two alleles for purple, it will have purple flowers. To have white flowers, it must have two alleles for white.

Mendel worked out that each individual has two alleles, but when they breed each passes on only one. Which of the two is passed on by each parent is a matter of chance. Mendel assumed that the alleles were split at the point when the germ cells (egg and sperm cells) are produced.

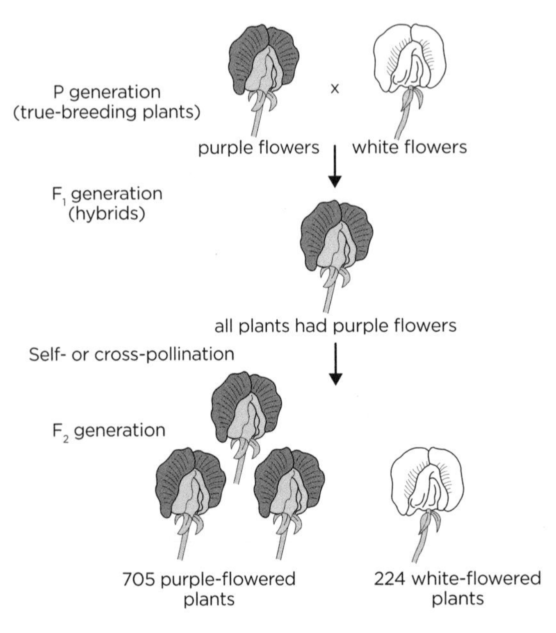

Mendel discovered that while purple was dominant in the hybridization of white and purple pea plants, surprises could occur.

This model explained his findings – a quarter of the offspring of two purple-flowered plants could have white flowers because by chance they had inherited white-flower alleles from both parents, who each had one allele for white flowers and one for purple flowers. Unfortunately, although Mendel published his research in 1866, it was neglected until 1900.

Stringy things

Even before Mendel began experimenting with his peas, the mechanism of inheritance had been seen, but went unrecognized. The Swiss botanist Karl von Nägeli saw a tangled network of stringy structures inside the cell's nucleus in 1842. A Belgian zoologist, Edouard van Beneden, noticed in 1883 that the stringy things separated into two bunches when germ cells are formed, so that each has half the number of chromosomes found in a normal body cell (though they weren't called chromosomes at this point). Five years later, in 1888, Beneden gave the stringy things that name.

Chromosomes untangled

The theoretical scientist August Weismann made the link between heredity and chromosomes in the 1880s before they had been named. Like Mendel, he suggested that germ cells contain half the amount of genetic information of normal body cells, and that when two germ cells (egg and sperm) come together they therefore provide the right amount for a new organism. Because the

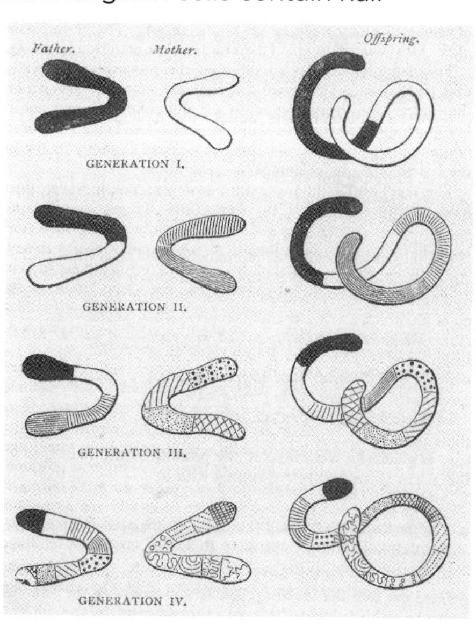

Weismann's illustration of heredity shows the two components from the parents on the left and the resultant mix in the offspring, right. Each row represents a new generation; the first parent is the offspring from the previous row.

new organism has a mix of genetic information from both parents, this provides a means of inheritance that can support the theory of evolution and reflect everyday observations. Further, he pointed out that the mechanism of inheritance was chemical – there was nothing particularly mysterious about it.

Putting the pieces together

Mendel's findings were rediscovered in 1900 – three times, independently – and it all started to fall into place. In 1903 an American, Walter Sutton, and a German, Theodor Boveri, both suggested that paired chromosomes are the means by which inheritance works, as described by Mendel. The theory is often known as the Boveri-Sutton chromosome theory. They also suggested, for the first time, that the chromosomes are all different so they carry different characteristics.

> ***'Heredity is brought about by the transference from one generation to another of a substance with a definite chemical, and above all molecular, constitution.'***
> August Weismann, 1885

> ***'The association of paternal and maternal chromosomes in pairs and their subsequent separation during the reducing division . . . may constitute the physical basis of the Mendelian law of heredity.'***
> Walter Sutton, 1903

From chromosomes to genes

It soon became clear that the chromosome is too big to be the basic unit of heredity. A human has 23 pairs of chromosomes, but we have far more than 23 variable characteristics. The answer to this is that the chromosome is divided into smaller units; these were named genes in 1909. The first assumption was that characteristics/genes are inherited in bunches, with all those on one chromosome going together. But that didn't produce enough variety, either.

The next breakthrough came in a room full of flies in 1910. Thomas Morgan had set up a research laboratory at the University of Columbia where he and his colleagues worked with fruit flies. They found that while some characteristics are often inherited together, it is not always the case. Morgan suggested that chromosomes can break and switch parts with another. This

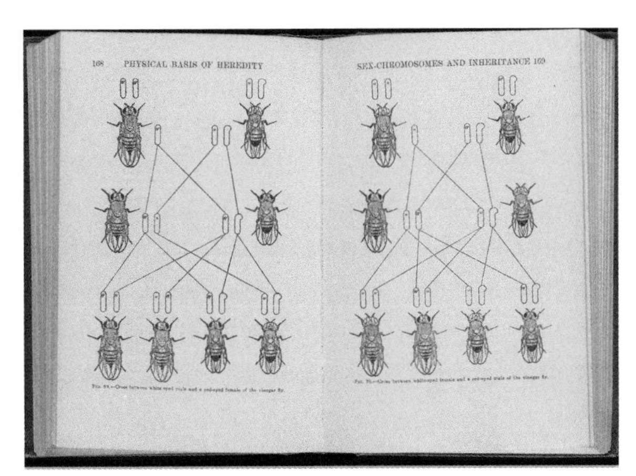
168 PHYSICAL BASIS OF HEREDITY

SEX-CHROMOSOMES AND INHERITANCE 169

Morgan's illustration of the inheritance of eye colour in flies, 1919.

process is called chromosomal crossover. Genes on the same 'chunk' of chromosome are inherited as a group. There is a greater likelihood of a break falling between genes that are far apart, so they will be inherited separately more often than genes that are close together on a chromosome.

Statistical analysis of how frequently different characteristics were inherited together in his flies led Morgan, with his student Arthur Sturtevant, to work out the relative positions of the genes and draw the first gene map in 1911.

DNA does it

Even then, no one knew how genes worked or what they were made of.

DNA (deoxyribonucleic acid) was first found in 1869 (in combination with RNA) and named 'nuclein' because it was found in the nucleus of cells. Its chemical composition emerged slowly until in 1919 it was clear that DNA is a long, stringy molecule made up of a series of chemical units called 'bases' linked together with phosphate groups. No one thought it likely that DNA could be the means of inheritance, though.

It took until the 1940s to find that DNA is the material of genes. Then, in 1944, Oswald Avery was able to show that the characteristics of a bacterium could be changed by putting into it DNA from another bacterium.

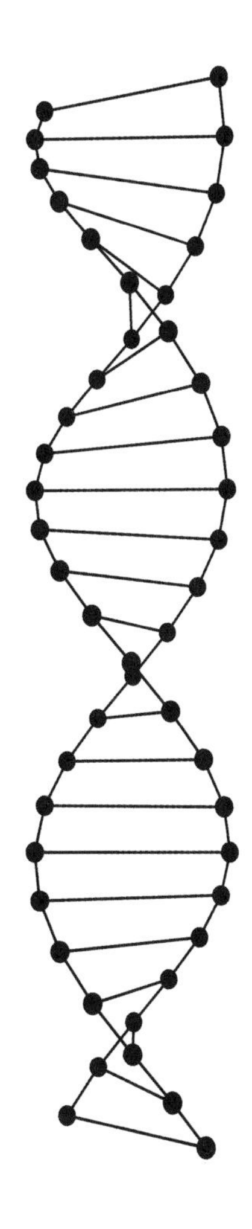

Unwinding the helix

The chemical composition of DNA was known, but not its structure, and it was its structure that held the clue to exactly how inheritance works. This was revealed in 1953 by three scientists working together in England: Francis Crick, James Watson and Rosalind Franklin. They found that DNA looks like a twisted ladder. Each side of the ladder is made up of a string of sugar and phosphate groups and each rung consists of two bases. The bases always occur in the same pairs: adenine with thymine and guanine with cytosine.

> ***'It has not escaped our notice that the specific pairing we have postulated immediately suggests a possible copying mechanism for the genetic material.'***
> Francis Crick and James Watson, 1953

Because the bases always occur in the same pairs, the DNA molecule can split down the middle, dividing the paired bases, and each half can replicate the other by producing the matching pair to complete each rung and adding another sugar-phosphate backbone. Crick and Watson concluded that the sequence of base pairs forms the genetic code. Chunks of DNA form genes, the entire molecule is a chromosome, and the order of base pairs within a gene codes information. The problem of inheritance was effectively solved.

Building bodies from genes

But that wasn't quite the end. There had to be a mechanism for turning the information contained in DNA into living bodies. Watson summarized it in a simple sequence:

DNA → RNA → protein

He meant that DNA is 'read' by a similar chemical, RNA, which then organizes the building of the proteins that govern life processes and build living organisms. Each gene holds the code for a single protein. Other information held on the DNA – it's not all genes – gives instructions about whether a gene will be 'expressed' (turned on). Together, it's like a full instruction set to build an organism, saying which proteins must be built where (and how) and when.

That RNA reads information from DNA and arranges the building of a protein was proven correct in 1964. Proteins are made from separate units called amino acids and 20 amino acids are arranged in various combinations to make the proteins used in the human body. It takes four DNA base pairs to identify each amino acid; the sequence of base pairs in a gene lists the amino acids needed for the protein the gene codes for. We could think of each gene as a page in the recipe book for an organism. It's quite a long book: a human being is described by around 20,000 genes.

JUST LIKE MUMMY

Most of this discussion relates to organisms which reproduce sexually - they take genetic material from two parents. Simple organisms reproduce asexually; other, more sophisticated organisms can do this as well. They still have genes, and their characteristics are still coded in DNA, but they start from only one set of genes so Mendel's theory of inheritance does not apply. Instead, all the genes of a single parent are copied into a new organism. The result is a perfect copy of the only parent - a clone.

Organisms that reproduce like this have less opportunity to diversify. They must rely on random mutation to change their genes.

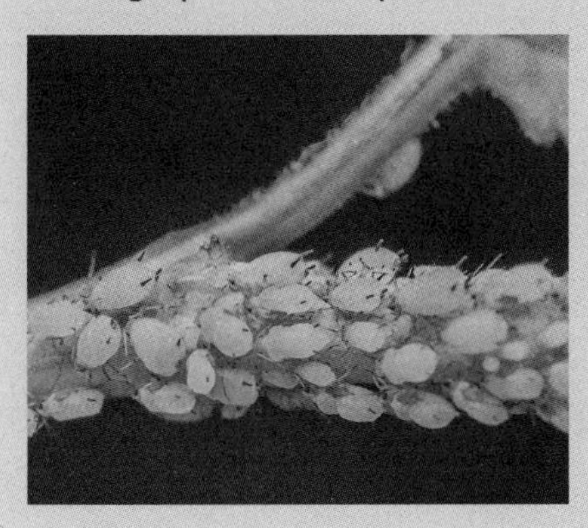

Aphids can reproduce sexually, but also produce many generations of clones.

Chapter 18
Information Theory

We are familiar with the idea of storing 'information', and that the information in a video is somehow 'larger' than a text message. But why is it? How do we measure the size of information?

THE IDEA

Information of all types can be measured by reducing it to a series of yes/no questions. It can be precisely quantified by working out the number of 'bits' it takes to encode it.

Introducing information

Any form of communication is a way of sharing information. The information can be of any type, from what you had for breakfast to a picture of a triceratops. It doesn't even have to be true – it can be a fairy tale, a poem or an outright lie.

Medium and message

Information is conveyed using a medium; the message and the medium are distinct. The medium could be language communicated as a song, a book, a prayer, a radio broadcast or a web page. It could be visual information, such as a painting, a YouTube video, an animated cartoon or a pie chart. It could be sound: a speech, a piano sonata or whale song. Some types of message can be conveyed only in certain media. You couldn't present a picture using a flute, or a dance using smoke signals.

For millennia, no one really needed to think very hard about information per se. It was conveyed in speech and in physical media, such as books and pictures. But then we began to use devices to transmit information over longer distances, first telegraph, then radio, telephone, television and eventually the internet. At that point, we had to start thinking about exactly how to communicate information and, as we wanted to send more and more of it, how to move it around more efficiently.

Cave paintings are an early example of people communicating information.

Coding and recoding

Information can't be transmitted without being put into some kind of code; even spoken language is a code. Written language is more obviously code – we have to recognize that 't' stands for a particular sound, or that a pictogram represents a certain spoken word, and then know how the word itself corresponds to an idea or an object. The English language can code any word using just 26 letters. That's quite economical. Being able to reduce information to a compact code makes communication more efficient.

The earliest writing system developed in Sumer (Mesopotamia) around 3100BC.

Although using just 26 letters is efficient for writing, it's still quite a lot of options if you want to design some sort of long-distance communication system. If you were using lights to signal, you'd need a way of distinguishing between 26 different options – perhaps 26 colours, or lights in 26 different positions, one to represent each letter. Similarly, if you were using signals along wires, you'd need 26 different wires. To make transmission easier, people hit on the idea of coding the code. Instead of sending each letter directly, we send a code that represents a letter. The best codes for easy transmission are binary – they have only two states.

Morse code

Early telegraphy used a code for letters and numbers (symbols) that consisted of long and short electrical pulses. This was Morse code, first used for telegraphs in 1844.

Each dash is the electrical current maintained for three times as long as the brief dot.

There is a slight pause between characters to stop the recipient jumbling them up.

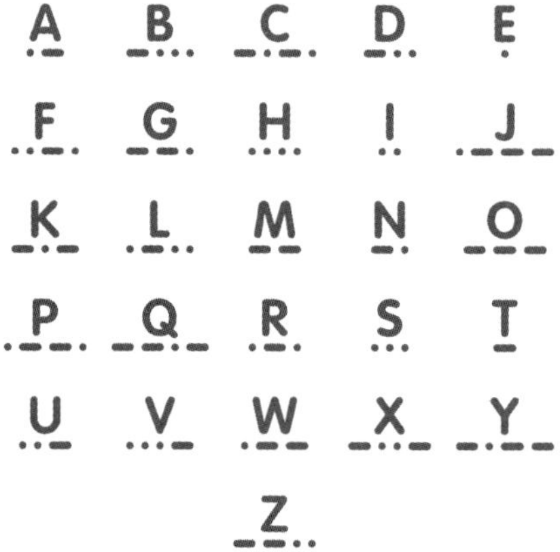

A •– B –••• C –•–• D –•• E •
F ••–• G ––• H •••• I •• J •–––
K –•– L •–•• M –– N –• O –––
P •––• Q ––•– R •–• S ••• T –
U ••– V •••– W •–– X –••– Y –•––
Z ––••

Being binary

Morse code might look at first glance as if it has three components: dots, dashes and

HORSE OR MORSE

When Morse code telegraphs were introduced, 90 per cent of messages were still carried by people on horseback – so it represented an incredible increase in speed of transmission!

silent spaces. In fact, it has only two: current-on and current-off. A dot was current-on briefly, and a dash was current-on for three times as long. Similarly, a short pause separated dots and dashes and a long pause separated symbols.

The Morse message:

. . . . . . - . . . -. . ---

means 'hello'. If we represent current-on by (=) and current-off by (.), the message looks like this and it's more clearly binary:

=.=.=.= . . . = . . . =.===.=.= . . . =.===.=.= . . . ===.===.===

If a code can be reduced to a binary form, there are many ways in which it can be transmitted and stored. Computers ultimately store (and transmit) information in binary form, coded as a string of 0s and 1s. These can correlate to current/no current, or +/- charge, or the presence or absence of a pit in the surface of a DVD, for instance.

Smoke signals are a code that uses a binary state – either there is a puff of smoke or there is not one.

Thinking about information

People only began to think about information as a measurable quantity when electronic communications became more important and widespread. Then it became necessary to compare information of different types in a way that had not been necessary or even possible before. For example, we might say intuitively that a picture 30cm (12in) square is larger than a picture 10cm (4in) square, but we could not say whether either picture was larger or smaller than a recording of a song – the question appears to be meaningless.

But it's not meaningless any longer. Information theory says that information is quantifiable and comparable in and of itself with no reference to its content. It provides a way of measuring information – in 'bits' – which allows us to compare information of all types to say how much capacity is needed to store it or to transmit it.

Asking questions

There are two important parameters which define any message. These are the total number of symbols (such as letters or digits) in the message, and the set of symbols from which one is selected at each point. If a message consists of a four-digit binary number, the number of symbols is four and the number of options at each step is two (0 or 1). We can draw a 'decision tree' to show all the possibilities.

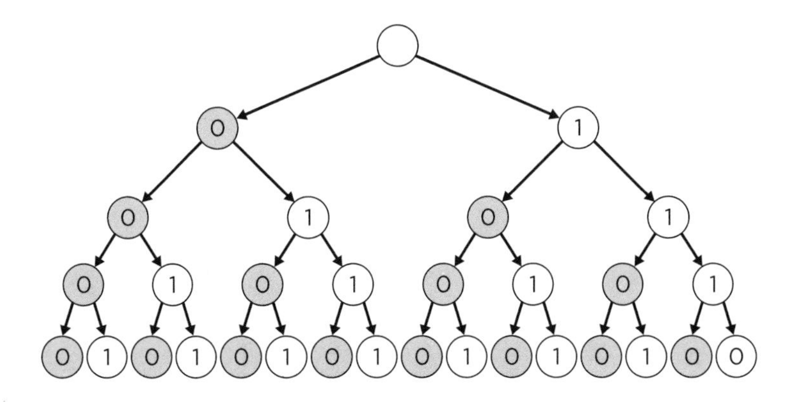

The tree shows a series of questions, which can be answered with 0 or 1 at each decision point. To see the possible messages, read along the branches: so the first possibility is (top down) 0000 and the next is 0001.

There are four decisions about which symbol to choose, and two symbols to choose from, 0 or 1, so there are:

$2^4 = 2 \times 2 \times 2 \times 2 = 16$ possible messages.

Most messages start off as something more than a string of 0s and 1s. To convert them to something we can store or transmit in a binary form, we need to interrogate the information. For a normal message in text, we would need to ask a string of questions about each letter that can each be answered with Yes or No and stored as 1 or 0. The most obvious way to do this is to ask 'Is it A?' then 'Is it B?' and so on – but that would take up to 25 questions for each letter of the message.

Asking fewer questions

Luckily, there is a way round this. Suppose we ask instead whether the letter is in the first half of the alphabet – is it less than M? The answer is either Yes or No (1 or 0). Then we split that half of the alphabet, and so on, until we have the letter. It will never take more than five questions to identify a letter, and never fewer than four (the average is 4.7).

- Is it less than M? – Y/N
- Is it less than G? – Y/N
- Is it less than D? – Y/N
- Is it A or B? – Y/N
- Is it A? – Y/N

Working it out

You could find out how many questions have to be asked by trial and error, but that's not efficient. In 1928, the American electronics researcher Ralph Hartley found the formula for working out how many questions would be needed to reduce any kind of message to binary form: to 0s and 1s or yes/no answers. The number of questions needed is:

$$n \times \log_2(s)$$

where 'n' is the number of symbols in the message and 's' is the number of options (so for the alphabet it's 26). Log_2 means 'logarithm in base-2' which can be looked up in a table or on a calculator; '$\log_2(s)$' is the power to which 2 must be raised to give the number 's'.

So if we wanted to transmit a message 10 characters long using an alphabet of 26 characters, the formula would be:

$10 \times \log_2(26)$

$\log_2(26)$ is 4.7 (because $2^{4.7} = 26$) – this is the number of questions needed to code one letter, remember. We need to ask the set of questions 10 times, so the answer is:

$10 \times 4.7 = 47$

This formula was revolutionary as it gave a method of quantifying information, measuring its size, in a way that was independent of the type of information. The principle can be extended to any information that can be interrogated and represented by a decision tree: a photograph, a document, a long number – anything.

> ***'What we have done then is to take as our practical measure of information the logarithm of the number of possible symbol sequences.'***
> Ralph Hartley, 'Transmission of Information', 1928

Messages aren't random

Asking questions about which symbol comes next works, but it's not very efficient. It takes no account of frequency or patterns in the sequence of letters or other symbols. Anyone who has played Scrabble will know that some letters are used more than others and some are used very seldom. The scores for different letters in Scrabble are related to how often they are used, with a high score attached to letters that are seldom used and hard to get rid of. Similarly, some patterns of letters recur (and some are never used).

CLAUDE SHANNON, 1916–2001

Shannon's fascination with communication started early. As a boy growing up in Michigan he built a telegraph line from fence-wire between his house and that of a friend. He went on to study mathematics and engineering and did a PhD in maths at MIT. His master's thesis showed how the mathematics of George Boole, known as Boolean algebra, could be implemented in electronics. All digital computing now relies on this work, using logic gates and the representation of true and false as 0 and 1 in circuits that carry out calculations and make decisions.

Shannon began working for Bell Laboratories in 1941 and was involved in cryptography work that was top secret. He was a conspicuous and slightly eccentric figure at Bell, where there was probably no shortage of eccentrics: he would travel through the corridors on a pogo stick or a unicycle.

A more efficient way of sending messages would use information about patterns and frequency to cut down the number of questions we need to ask. For example, in English 'e' is used more often than 'y', 'ion' is a common sequence, and the sequence 'qj' is rare (but must be allowed).

In the 1940s, American mathematician and electronics engineer Claude Shannon explored this non-randomness. He looked at how the decision about a symbol is influenced by previous choices of symbol in the sequence. Let's look at frequency first.

Entropy

In a completely random series of events, such as coin-tosses or rolling a die, there is maximum unpredictability. Shannon called this unpredictability, or surprise value, 'entropy'. But in many types of message, including written texts of all types, the sequence is a bit more predictable. This means the entropy is reduced. Entropy had already been described in relation to liquids and gases around 70 years before Shannon used it to talk about information.

The predictable patterns of language can be used to cut down the number of questions we need to ask to find out what the next letter will be each time. Imagine a messaging system that uses four letters, A, B, C and D, but they don't all appear with the same regularity. A occurs 50 per cent of the time, B occurs 25 per cent of the time and

C and D both occur 12.5 per cent of the time. We could find out which letter comes next in a message by asking two questions. The first would be, 'Is it A/B?' This tells us it's either A or B or it's C or D. The next question asks either 'Is it A?' or 'Is it C?' But there is a better way. We can use the different probabilities to reduce the average number of questions we need to ask.

As half the time the next letter will be A, it's most efficient to ask as the first question, 'Is it A?' Half the time we will get away with asking just this one question. The next question should be 'Is it B?', as this will sweep up 50 per cent of the remaining possibilities. Of course, a quarter of the time the letter will be C or D and then we have to ask three questions. But on average, we won't. The probabilities for the different numbers of questions are:

1 question (A) = 0.5
2 questions (B) = 0.25
3 questions (C or D) = 0.25

Adding the probabilities together, the probable number of questions needed to determine any one symbol is then:

$$(1 \times 0.5) + (2 \times 0.25) + (3 \times 0.25) = 1.75$$

As the answers to the Yes/No questions are stored as binary digits, Shannon called the storage unit a 'bit' (a contraction of **b**inary and dig**it**). So the amount of information (number of bits needed to store the result) is 1.75 for each symbol. If we have a

'word' that is four symbols long, the total number of bits will be 4 x 1.75 = 7 bits.

Shannon put this into a formula for entropy:

Σ (p x number of questions)

which means add up (shown by sigma, Σ) the probability for each symbol (p) multiplied by the number of questions. With our example, it means the calculation we have just done:

(1 x 0.5) + (2 x 0.25) + (3 x 0.25) = 1.75

With a bit of mathematical jiggling around, he produced an equation for defining entropy:

entropy = Σ p x $\log_2(1/p)$

DOLPHINS, ALIENS AND ENTROPY

Beginning in 1961, researchers interested in searching for extra-terrestrial intelligence by scanning radio signals from space began to study dolphin communications. The researchers discovered that baby dolphins, just like human babies, make nonsensical sounds called 'babbling'. As a human baby becomes accustomed to its native language, the range and pattern of babbling changes, narrowing as it moves towards normal speech. A graph plotting how frequently each sound occurs is pretty even for a baby. A similar graph drawn for an adult speaker has about a 45 degree slope as some sounds are used very often in a language, others less so and some hardly at all. In terms of Shannon's theory, entropy decreases as language ability emerges.

The researchers found the same pattern in dolphins, with the range of utterances reducing as the dolphins mature and tending towards a 45 degree slope in adulthood. This suggests that as the babies learn to speak dolphin, the less useful sounds in their language occur less frequently. The researchers concluded that this pattern might represent a universal structure of languages – perhaps even languages which don't originate on Earth. If we found the same pattern in a signal from outer space, it might be from an intelligent being sending a message.

This formula for entropy in information is fundamental to all modern communications, to computer storage and to a host of other applications. It gives a way of measuring information of different types and comparing the size of, say, a recording of a speech and a scanned painting. The type of information is not important, nor does the meaning matter; as Shannon said, meaning is 'irrelevant to the engineering problem'.

What is size?

'Size' becomes hard to define once we start reducing information to binary and coding it in different ways. Whether a picture 30cm (12in) square is bigger than one 10cm (4in) square depends on the questions we ask when encoding it. In an extravagant and simple coding system that asks of each pixel what colour it is, the 30cm picture will always be bigger as we have to ask the same questions of a larger number of pixels. But if we use a more sophisticated method, such as the ones used by modern graphics programs, the number of questions reduces. We can include the question 'Is this pixel the same colour as the last?' – it often will be. That cuts out a lot of questions. Using this type of method, a very ornate and detailed picture only 10cm square takes more information to code – so is larger in terms of information – than a very simple picture made of blocks of solid colour that is 30cm square.

Chapter 19

RNA World Theory

The theory of evolution provides a model for how we – and everything else – got here. But where did the earliest life forms come from?

THE IDEA

Before the DNA-based life forms that now populate the world, very simple early life forms developed around self-replicating RNA.

RNA world

All living organisms carry genetic information which acts as a recipe or blueprint for the organism, coded in DNA (see page 234). When the organism reproduces, the genetic information is copied, so traits are passed from one generation to the next.

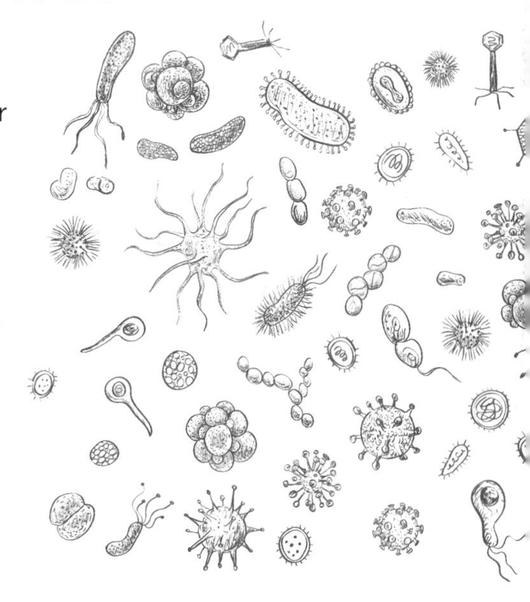

Viruses, which are on the border between living and non-living things, sometimes use a different chemical to code information about their make-up; this is similar to DNA, and is called RNA. It's possible that the ability to store and copy information in the form of RNA represents an early step in the evolution of life. The 'RNA world' theory suggests that molecules of RNA first copied each other and themselves, providing the starting point for self-replicating life to evolve.

It's all proteins in the end

All genetic information is coded as a sequence of recipes for making proteins. All your physical and physiological features are produced by the mix of proteins your body makes or fails to make in different parts. And every function your body carries out is controlled by the synthesis (building) of proteins.

While DNA holds the recipes for the proteins, it can't make them. The process of making proteins is extremely complicated. The recipe is 'read' by another chemical, RNA, which enables the production of the proteins. This complex procedure can't have evolved in a single step. The RNA world theory suggests that RNA originally carried out both major steps itself: it held the instructions and catalysed (facilitated) making the proteins.

Life needs to multiply

The crucial feature of life is that it is self-replicating: it can make copies of itself from chemical building blocks. We don't know exactly how life started on Earth but it seems likely that first of all there were chemicals that could copy themselves and then more complex systems built up from these.

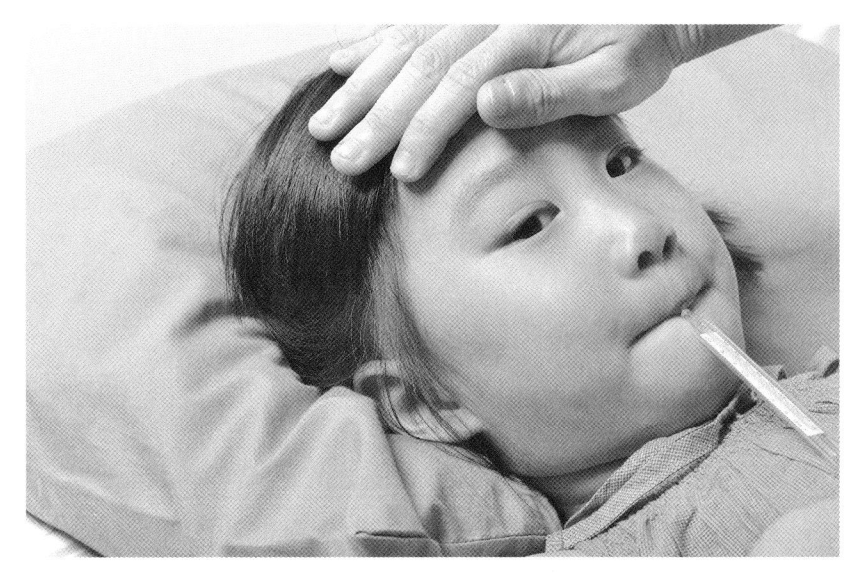

Human diseases caused by RNA viruses include the common cold.

The most important chemicals in modern life forms are polypeptides, including proteins (see page 266). These are built up from small chemical groups called amino acids which link together. It's possible for amino acids to be produced in conditions that existed on early Earth and they have even been found on rocks from outer space, which suggests they form quite easily. Even so, no one has succeeded in prompting polypeptides to build copies of themselves.

Copying from blocks

A different type of chemical grouping can produce copies of itself, though. Polynucleotides (see box) such as RNA can self-replicate, but the process is slow and prone to error and polynucleotides really need a catalyst. Then in 1982 it became clear that RNA not only self-replicates, it can also act as a catalyst in the reaction: it can be its own catalyst. This is a bit like being able to pick yourself up by your own ears. It's a pretty cool trick. It's an even cooler trick if being able to pick yourself up by your ears leads to the evolution of all life on Earth.

It's not clear whether RNA sprang up as the first self-replicating molecule. It's rather long and complex for that. It is possible –

BUILDING FROM BLOCKS

A polypeptide is made of connected amino acids. Proteins are made up of one or more polypeptide molecules. They start as a long chain and then coil and fold into a complicated shape. The shape is essential to the chemical behaviour of the protein. The word 'polypeptide' is often just used to mean polypeptides smaller than proteins.

A polynucleotide is made of linked nucleotides. A nucleotide has three parts: a sugar, a nitrogenous base and a phosphate group. It's built with links between the sugar and phosphate with the bases sticking out. RNA has a single strand of sugar/phosphate groups. DNA has two strands linked by their bases and twisted into a helix.

assuming the theory is correct at all – that there was a pre-RNA world that was ruled by a simpler polymer. Some types of RNA may be good at activities that encourage the copying of RNA – perhaps speeding up the rate at which the components of RNA form, for example. It's easy to see that a population could form of RNAs which work together, providing a healthy chemical environment for their own replication.

While all the different types of RNA are floating around freely, the replicases would randomly copy whichever

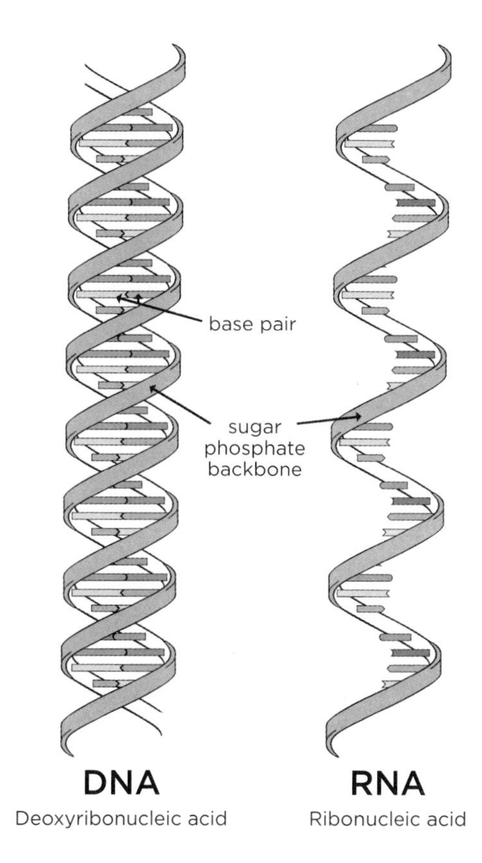

While DNA forms a double helix linked by paired bases, RNA is a single spiral with bases sticking out.

other bits they came across. Sometimes these would be other replicases and sometimes they would not. For evolution, we need lots of replicating of the right types of RNA to happen, so this randomness is not very efficient. That's where a container comes in handy . . .

MAKING RNA

Recent experiments have shown that strings of RNA can form from component chemicals in the right conditions, such as those that were available on early Earth. Components of RNA brought close together form bonds; this has happened when components have come together on the surface of clay particles or have become concentrated in freezing salt water. Many different types of RNA could have emerged in this way. Some would have been replicases, capable of copying other RNA strings.

Keeping the inside in and the outside out

A set of self-replicating molecules is a good start, but it's not life. Before any more progress could be made, some kind of container or envelope was needed to separate the important bits of RNA from other bits. In modern cells, the inside is kept in and the outside is kept out by a membrane. It's likely that a simpler type of mechanism developed in the early days of the RNA world.

Fatty acids are compounds with long molecules. The two ends are very different: one end is attracted to water (hydrophilic) and

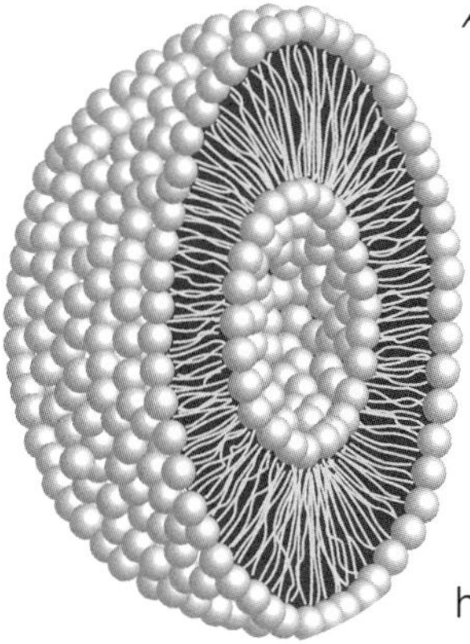

A vesicle forms a hollow ball with the water-loving 'heads' of the molecules forming the inner and outer layers.

the other end is repelled by water (hydrophobic). The best way for these molecules to cope with being in a watery environment is to gang together and form clusters with all the hydrophobic tails in the centre and the hydrophilic heads on the outside. A simple cluster of one layer is called a micelle.

If a whole bunch of micelles get together, they tend to rearrange into a double layer which forms a ball called a vesicle. This has a space in the middle, with hydrophilic heads on both the inside and outside surfaces of the 'wall' of the vesicle and all the hydrophobic tails hiding within the wall. The vesicle forms a container that can float around in the water, but keep the water outside from mixing with that inside.

Making protocells

Fatty acids can form around hydrothermal vents (undersea spouts of hot water) without any chemical encouragement (catalysis). Another place they will form is on the same type of clay that helps

RNA to form – a lucky coincidence for evolution. They are in low concentrations to start with but if they were stranded in isolated pools there would be enough in a small area for them to clump into vesicles as water evaporates and brings them closer together.

When a vesicle forms, water is trapped inside. If there are pieces of RNA inside, including replicases, and the necessary components for building more RNA, they will get on and replicate. Occasionally, they might contain a population of RNAs that work well together.

RNA world theory suggests that RNA provided the starting point for self-replicating life to evolve.

At the same time, as more micelles collide with the vesicle, they will be incorporated into the structure. As the vesicle grows larger, it becomes more fragile and eventually breaks in half. It will tend to split into two vesicles which rapidly close their walls again. The RNA strands will be randomly split between the two new vesicles, but it's likely there will be some in each. *Voilà* – protocells that can reproduce enclosed within a membrane!

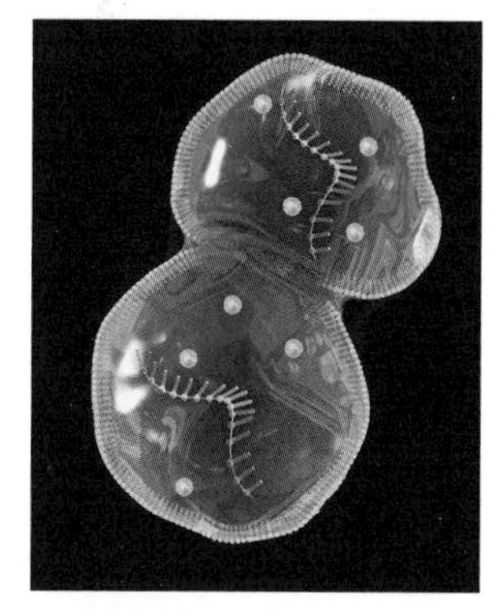

Could life have begun like this? A cell splits in two, with an RNA strand in each part.

Evolving and growing

Over time, there will be slight errors in copying and some RNA strings will fare better than others – perhaps they replicate more readily or quickly, perhaps they also encourage vesicle growth in some way – and these will start to predominate in the local population. Soon there will be an evolving chemical population. It's well on the way to being something we could call life. At some point, DNA would have taken over from RNA as the chemical that held the coding, with RNA retained to carry out replication. At this point, we would be out of the RNA world and into the beginnings of life as we know it.

Chapter 20

Atomic Theory

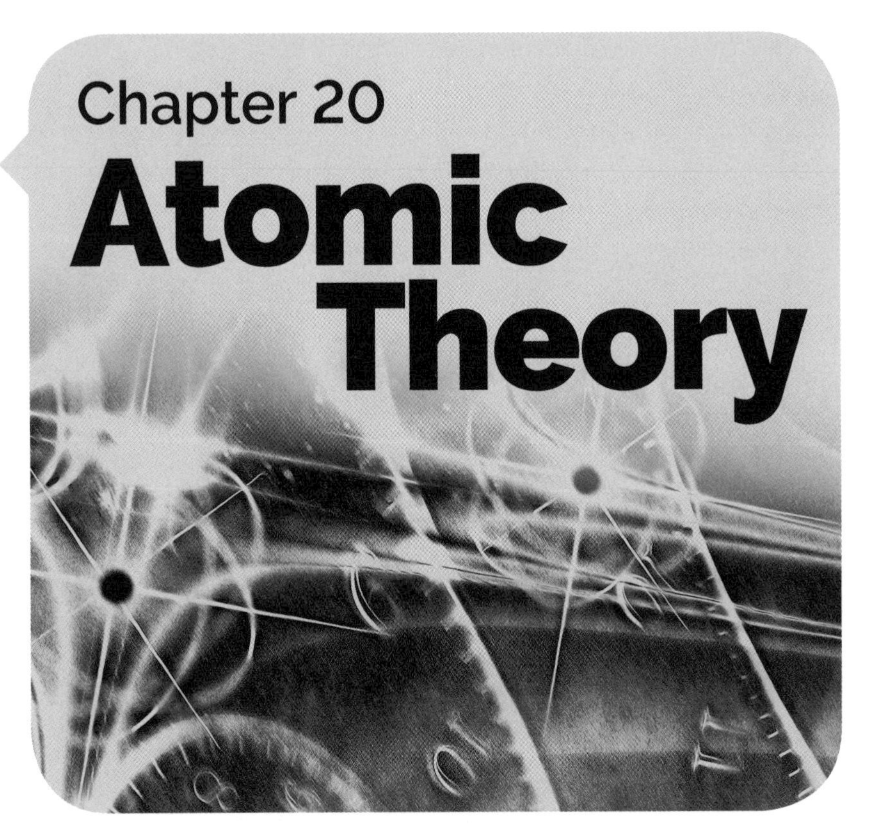

The idea that everything in the universe is made of tiny particles is very old. It has moved from the realm of philosophy into mainstream science.

THE IDEA

All matter is made of very tiny particles called atoms. The atom is the smallest particle that is specific to a particular type of matter (one of the chemical elements).

All or nothing?

There are two possible models for matter and indeed the entire universe: either it is a mass of continuous matter, or it is made up of separate tiny chunks. If the latter, there must be empty space between them. The empty space was a problem for the Ancient Greeks, some of whom were so troubled by the idea that there might any kind of nothingness that they assumed the universe was jam-packed full of matter.

But if matter is continuous, movement becomes hard to explain. For something to move, there must be a space for it to move into. If there

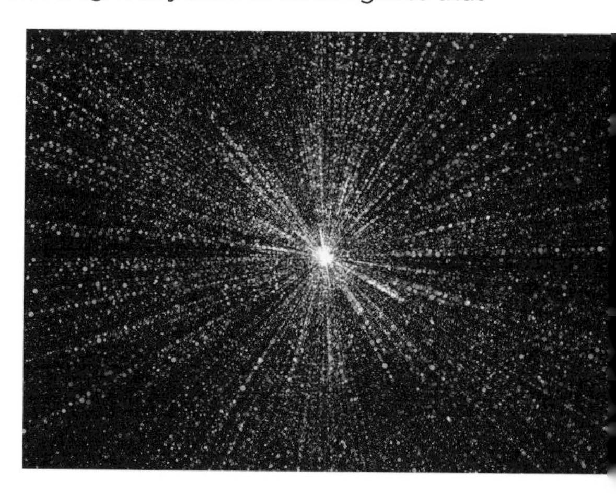

Only if we accept that there can be empty space do particles become possible.

is no space, it must push something else aside. But where does the something else go? One solution - which looks rather odd to us - is that all movement and change are illusions. The Greek philosopher Parmenides argued that all material is a single, continuous, unchanging mass and there is no void. The alternative view - that matter comprises an almost infinite number of particles existing in a void - was proposed around the same time.

First atoms

The idea that matter is made up of tiny particles was first suggested in Ancient Greece in the 5th century BC and in India in the 4th century BC. In Greece, Leucippus and his pupil Democritus proposed that all matter comprises tiny indivisible particles which they called 'atoms'. They said

Democritus proposed that all matter is produced by the interaction of atoms in space.

that all the different types of matter and conditions in the world are produced by the interaction of atoms in space and in their different arrangements and combinations. It was a pretty good guess, but the idea didn't really take off. The influential philosopher Aristotle did not think there could be a void, and without a void there can't be atoms. Although the idea resurfaced every now and then, it did not really come back with any vigour for 2,000 years.

From atoms to corpuscles

Galileo Galilei.

From around 1600, more scientists began to entertain the idea of something at least a bit like atoms. Those who were persuaded included Galileo Galilei, Pierre Gassendi, Isaac Newton and Robert Boyle. These particles were called 'corpuscles' (meaning 'little bodies'). Unlike the atoms of Democritus, they were not necessarily considered indivisible. For instance, it was believed that corpuscles of the liquid metal mercury could permeate corpuscles of other metals and change their properties. This was fundamental to the alchemical belief that base metals could be converted into gold.

Atoms and elements

The Ancient Greeks had believed that all matter is composed of a mix of four 'elements' – fire, air, water and earth. Until the beginnings of modern chemistry and the recognition of the chemical elements there was not much progress to be made with exactly how atoms could work. The origins of modern atomic theory lie with the English chemist John Dalton, who published his work in 1805 and 1808. Dalton's atomic theory still forms the basis of the modern understanding of atoms, though there have been some adjustments. Dalton's four principal findings are:

1. All matter is made of atoms, which are indivisible and indestructible.
2. All atoms of a given element are identical in terms of both their mass and their properties.
3. Compounds are formed by combining two or more different kinds of atoms (that is, atoms of two or more elements).
4. A chemical reaction is a process that rearranges atoms.

Dalton began with the principles of conservation of mass and constant composition. Conservation of mass means that in a closed system (one which doesn't allow anything to be added from the outside or taken away) matter can't be either created or annihilated – the mass of matter will remain the same, even though it may change in form. For instance, if you melt an ice cube (a physical change), you will have the same mass of water as of ice though it will be a different shape and a different volume. If you seal a sheet of paper in a box and burn it (a chemical change), the mass of the contents (ash and gases) will be the same as the mass of the original contents (paper and gases). The law of constant composition means that a chemical compound will always be made of the same ingredients in the same proportions – so pure water will always have the same chemical composition, as will salt (sodium chloride), and any other compound.

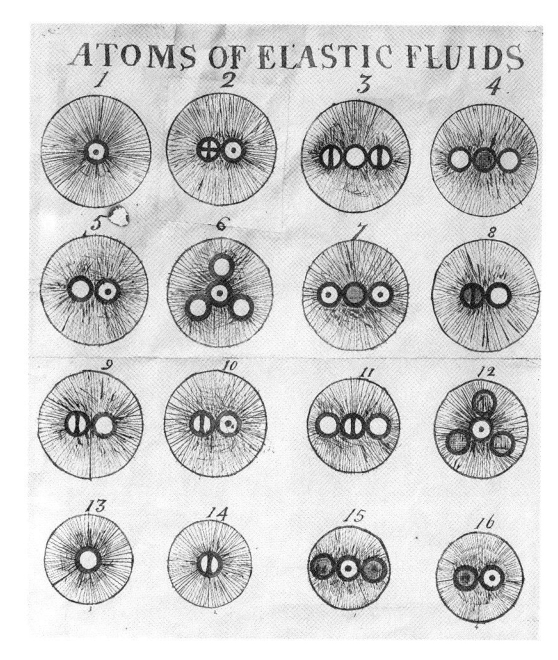

Dalton's drawing shows the molecular make-up of different gases ('elastic fluids'). Each design of spot represents an atom of a different element.

Dalton proposed that the conservation of mass and the law of constant composition could both be explained by matter being

composed of tiny, indivisible particles – or atoms. He considered an atom to be a 'solid, massy, hard, impenetrable, movable particle'.

He stated that every atom of an element is identical, but the atoms of each element differ from those of every other element. This gives the elements their different properties. Atoms always combine in the same proportions to make the same compound. As atoms are indivisible, it's always possible to find a whole-number ratio of the different atoms in a compound. When they are combined, atoms are not changed. They can be rearranged to make new chemical compounds, but no atom is destroyed, created or altered by the process. It's rather like making a model from coloured pieces of LEGO, then taking it apart and making another model with the same bricks. No brick is changed – it stays the same size and shape – but what you have made could be very different.

Divisible after all

Although Dalton's theory is the basis of modern atomic theory, his assertion that atoms are indivisible particles is no longer considered true. What is true is that an atom is the smallest particle that can be called a substance. Atoms are made up of subatomic particles – very tiny components – and the atoms of different elements differ only in how many subatomic particles they contain. If you could actually cut an atom in half, so that it had half the number of subatomic particles (assuming it started with an even number of each) the result would be two atoms of a different element.

In the late 19th century, scientists discovered that if they ran an electric current through a gas in a glass tube it produced a stream of particles. With the inside of one end of the tube painted with

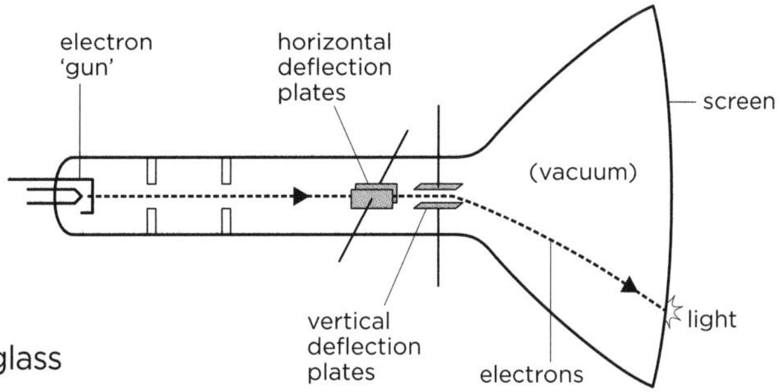

A spark of light is produced as electrons are fired in a cathode ray tube.

phosphor, a spark of light was emitted when the beam hit it. The English physicist J. J. Thomson investigated discharge tubes further and discovered that the 'rays' emitted were actually streams of very tiny 'corpuscles' with a negative electrical charge – we now call them electrons. He found their mass to be about a two-thousandth the mass of a hydrogen atom. This meant the rays must be composed of particles smaller than an atom – so the atom wasn't the smallest particle after all.

Thomson found that if he changed the material used for the cathode (which was the source of the particles) the nature of the 'ray' or emitted particles was the same. From this he concluded that the particles are the same in every type of atom.

Atoms don't have an overall negative charge, so Thomson realized that the electrons must be counterbalanced by a positive charge in the atom. He constructed a model of the atom in which he assumed that the negative and positive charges were pretty equally distributed. He suggested a positively charged matrix, or sea of charge, within which the negative particles milled around. He named it the 'plum pudding' model – he envisaged the positive matrix as a spherical pudding, and the free-floating electrons as raisins in the pudding.

Puddings and protons

The next step was taken by Ernest Rutherford, a chemist from New Zealand working in England. He bombarded a thin sheet of gold foil with alpha particles. (He used gold because it can be beaten very thin, making a layer that is only a few atoms thick.) Rutherford didn't know what alpha particles were, except that they were very small, were produced by the radioactive decay of radium and had a positive charge. In fact, an alpha particle is identical to the nucleus of a helium atom; it has two protons and two neutrons but no electrons.

Rutherford's remarkable discovery meant the model of the atom had to be revised to include a lot of empty space.

Rutherford expected that the alpha particles would go straight through the gold foil, since he assumed the smeared-out positive charge of the plum pudding model would be too diffuse to deflect them. But he found that while most went straight through the foil, some were deflected a little on their path and a few bounced back. It was an astonishing result – it meant the plum pudding model must be wrong.

Rutherford concluded that a positive charge in an atom is concentrated into a particular, small area. Only when an alpha particle collided with this area was there sufficient repulsive charge to send it back the way it had come. He called the area where the positive charge was concentrated the 'nucleus'. It was also obvious that for most of the particles to go straight through, there must be a lot of empty space in matter – there were big gaps between the nuclei.

> ***'It was quite the most incredible event that has ever happened to me in my life. It was almost as incredible as if you fired a 15-inch shell at a piece of tissue paper and it came back and hit you.'***
> Ernest Rutherford

Planetary and non-planetary atoms

Rutherford proposed a new model of the atom so that it was something like a planetary system. It has a tiny nucleus where all the positive charge – and the mass – of the atom is concentrated,

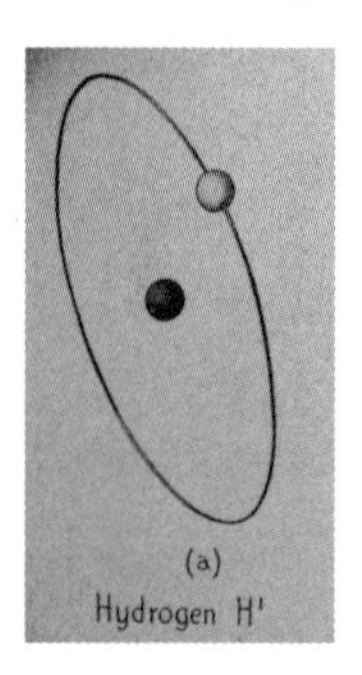

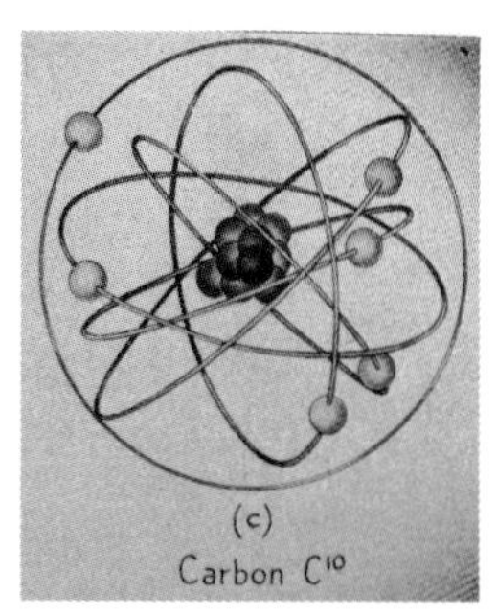

Atoms are like tiny solar systems with electron planets revolving round protons and neutrons.

surrounded by a set of orbiting electrons. He discovered the particles that carry the positive charge in 1911, calling them protons. The notion of the nucleus raised new problems – such as why the electrons were not attracted by the positive nucleus sufficiently to collapse into it.

The Danish physicist Niels Bohr, who worked with Rutherford, proposed that the electrons don't just mill around the nucleus willy-nilly picking any orbit they like. Instead, he suggested, they inhabit very specific levels. This model could explain the phenomenon of emission spectra. When gases are heated, they produce light at very specific, characteristic wavelengths. Bohr explained this in terms of energy being emitted by the atoms as electrons jump between allowed levels (see pages 293–4). For an electron to move to an energy level closer to the nucleus, it needs to gain energy. As it moves to a level further from the nucleus, it releases energy. A set number of electrons can exist at each energy level, each in its own allocated patch of space, called an orbital.

Electrons are not 'allowed' to be anywhere except in one of the

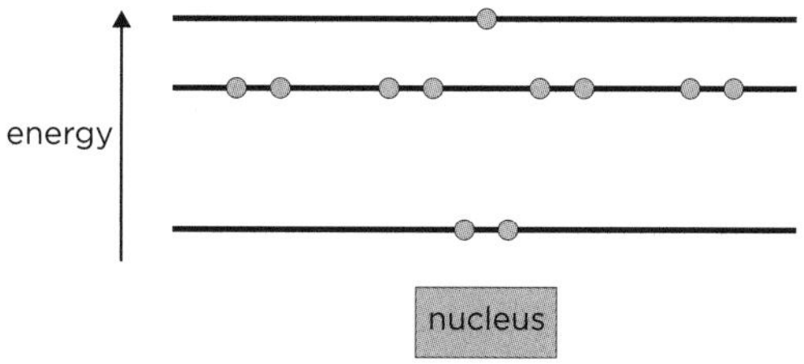

energy

nucleus

The orbitals of a sodium atom shown as straight-line energy levels.

prescribed orbitals, but where exactly they are within that can't be stated with certainty or predicted. The orbital simply describes an area of probability - the area where the electron is most likely to be is where we draw the orbital. Within that it is decidedly not orbiting the nucleus like a planet orbits the Sun. Bohr's model replaced Rutherford's planetary model.

s, p, d, f – shapes of orbital

Although you might have seen diagrams that show the configuration of electrons in an atom as a series of concentric circles with dots representing the electrons, these are misleading. At its simplest a hydrogen atom could be drawn as a nucleus with one circle around it and a single electron drawn on the line. This doesn't mean, though, that the single electron goes around the nucleus in a circle and won't be found anywhere but on that circular path. Rather, the diagram is a stylized cross-section. The hydrogen electron's orbital is really a spherical cloud around the nucleus. This is the volume of space in which the electron is most likely to be found at any particular moment. (To encompass all the places the electron *could* ever be found, the orbital would need to enclose the entire universe.)

Few orbitals are actually spherical, but those that are spherical are designated by the letter 's' (sensibly). The two innermost orbitals are both spherical, though the second, 2s, is a hollow sphere which

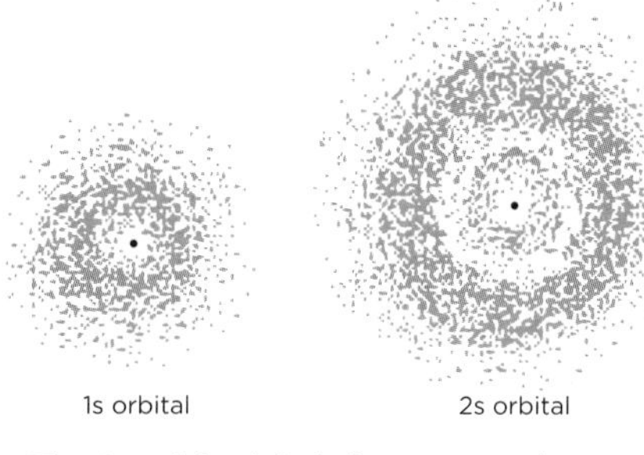

The two 's' orbitals form nested spheres.

encloses 1s. A more common shape is a 'p' orbital, which is rather like a dumbbell, or perhaps two balloons tied together at the nucleus.

There can be three 'p' orbitals at the same energy level, each at right angles to the other two. The second energy level of an atom such as sodium has one 's' orbital and three 'p' orbitals. Further levels have additional types of orbital, called 'd' and 'f' (there can be five 'd' and seven 'f' orbitals). Each orbital can hold two electrons. As an atom fills its energy levels, it adds one electron to each orbital in the outer level first, then adds a second to each. The 's' orbitals have slightly lower energy levels than the 'p' orbitals, so they fill first. The 'p' orbitals take one electron each in sequence, then a second one in the same sequence. They are followed by the 'd' orbitals, but these actually have a slightly higher energy level than the next level 's' orbitals so, for example, 4s is filled before 3d. This quirk only

A 'p' orbital has two lobes, one either side of the nucleus.

applies to working out the initial arrangement of electrons in an atom. When the atom forms bonds, it always fills orbitals in the order of levels (so 3d would come before 4s).

Elements and electrons

It might seem as though the order in which an atom orders its electrons is rather esoteric and of no real interest to anyone, but it is crucially important in terms of how reactive an element is and what types of compound it will tend to form.

Dalton stated that every atom of an element is identical. Although we now vary that slightly as isotopes have different numbers of neutrons, it is generally the case that every atom of a particular element has the

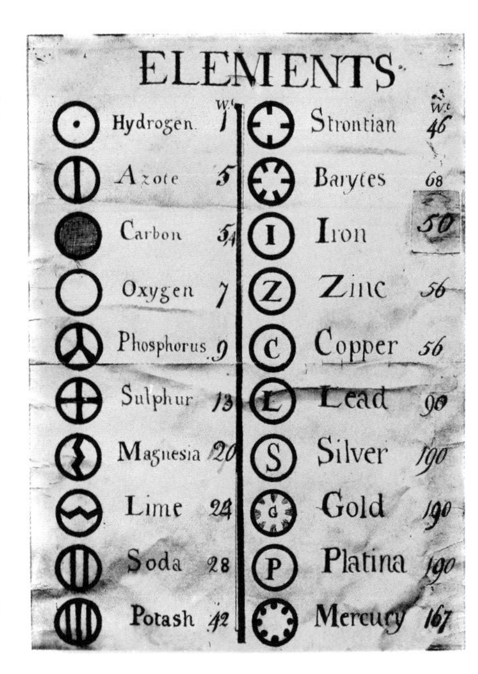

The symbols used for the known elements, before chemists began to use alphabetic symbols for them.

same number of electrons and protons, and the electrons are found in the same specific energy levels. Hydrogen, the simplest atom, has a single electron. Heavier elements have more electrons, all arranged in their energy levels.

There is an allowed maximum number of electrons at each energy level, and atoms are at their most stable when they have a full complement of electrons in all the occupied levels. As electrons generally (with exceptions) occupy orbitals from the nucleus outwards, atoms tend to grow larger as they have more electrons. This also means that any gaps where more electrons would be allowed are towards the outer edge of the atom. (Not that there really is an edge, as there is no boundary beyond which the electrons cannot stray.)

Sharing electrons

When elements combine to form compounds, atoms gang together in molecules. The atoms are held together by shared electrons. Imagine atoms of sodium (Na) and chlorine (Cl). Sodium has 11 electrons arranged in these orbitals:

1s (two electrons), 2s (two electrons), 2p (six electrons), 3s (one electron)

This can be written:

$1s^2\ 2s^2\ 2p^6\ 3s^1$

Sodium

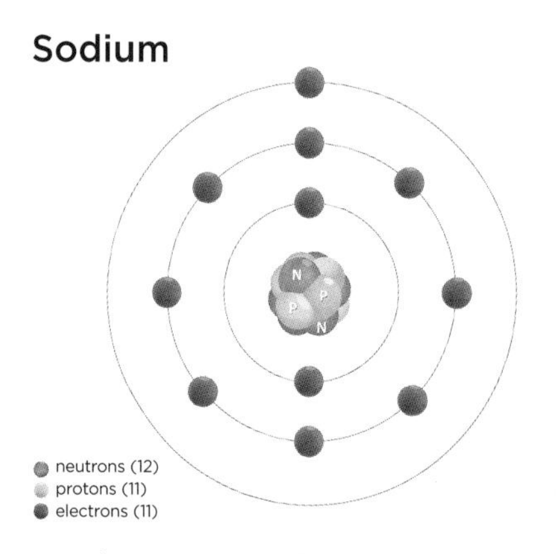

As there is only one electron in the 3s orbital, there is space for another. Sodium is unstable because of this single electron in the outer orbital (3). Sodium could become stable by getting rid of this spare electron.

Chlorine has 17 electrons, arranged as $1s^2\ 2s^2\ 2p^6\ 3s^2\ 3p^5$. To fill its outer orbital, 3p, it needs one more electron. There is an easy solution for sodium and chlorine – sodium can give away one of its electrons to chlorine. When this happens, two ions are formed. Sodium has lost an electron (with a negative charge) so then has a positive charge and is written Na+. Chlorine has gained an electron, so now overall has a negative charge and is written Cl-. The sodium and chlorine ions are attracted to each other because they have opposite charges, so they form a compound, sodium chloride (normal table salt). This type of bond, where one atom gives up one or more electrons to another, is called an ionic bond. Atoms tend to form ionic bonds that will give them a complete outer orbital, but the energy involved has to balance. If it takes more energy to pull another electron away than will be produced by making the compound, the reaction won't happen.

Atoms can also share electrons. Hydrogen has a single electron in its one orbital, but would be more stable with two. Two hydrogen atoms can share their electrons so that both, effectively, have two. This is called a covalent bond. Two hydrogen atoms together form molecular hydrogen, H_2. The shared electron(s) in a covalent bond set up a new orbital which overlaps both atoms. In more complicated atoms, there is a bit of jiggling about, moving electrons between levels and hybridizing orbitals to make them odd shapes. All this helps to maximize the energy efficiency of making bonds.

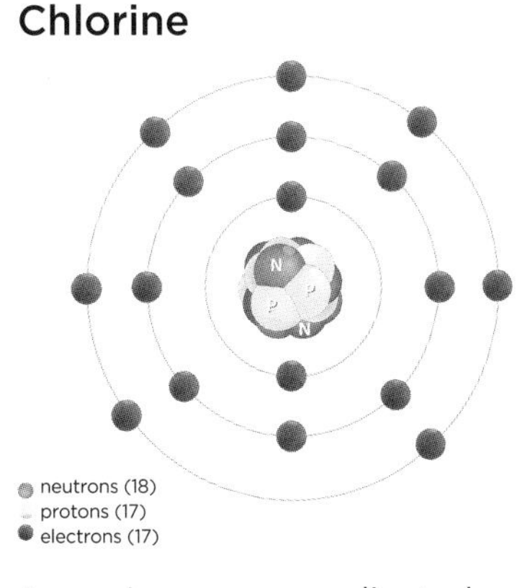

From atoms to all the universe

Modern atomic theory explains, through the arrangement and sharing of electrons, how elements combine together to form compounds. It can also explain the features of substances – why they are gases or solids at room temperature, for instance, or why they are hard or soft. It seems that Leucippus and Democritus were right (though for the wrong reasons): the shape of atoms really does determine how they group together and form all the matter in the universe.

Chapter 21

Quantum Theory

It might seem obvious that we know where things are. But, on the smallest level, perhaps we really don't know where anything is.

Bright lights and stripes

In the 1850s, the German chemist Robert Bunsen and physicist Gustav Kirchhoff built a spectroscope at the University of Heidelberg. Using a Bunsen burner, usefully just invented by Robert Bunsen, they heated substances and used their spectroscope to examine the light from the gases produced. They found that each produced a series of brightly coloured bands of light.

THE IDEA

Energy always comes in 'quanta' – tiny parcels of a fixed size. There are many consequences of this, including the way atoms can combine and the impossibility of making measurements without altering the state being measured. At the subatomic level, certainty breaks down and information about particles is replaced by probabilities.

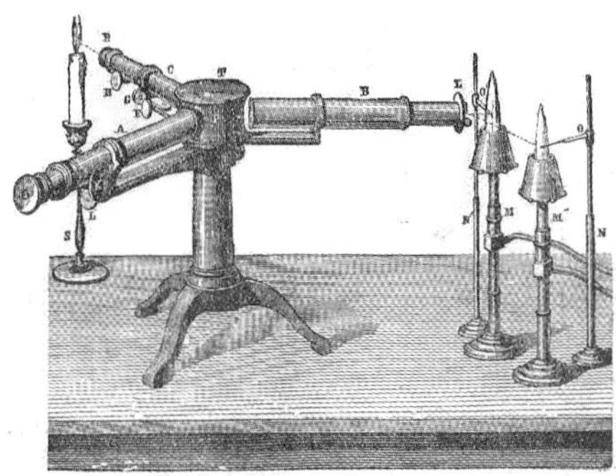

The Kirchoff-Bunsen spectroscope.

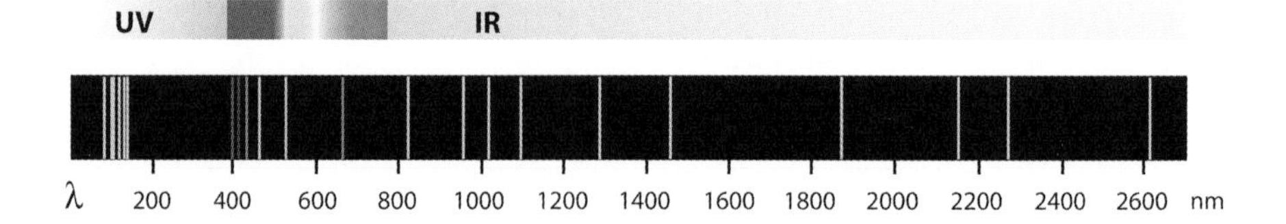

The spectral lines produced by hydrogen, from ultraviolet, through visible light and infrared.

The spacing and location of lines are different for each of the chemical elements, so an element can be identified from emission spectra. Bunsen and Kirchhoff used this phenomenon to build up a catalogue of spectral 'fingerprints' of the elements; but they didn't know why or how the elements emitted these precise bands of coloured light.

It was clear there was a pattern to the positions of the bands, but all attempts to link it to anything failed until 1885. Then the Swiss physicist Johann Balmer found a simple formula that gives the wavelength in most cases. The wavelength for the bands for hydrogen is always given by:

$$3645.6 \times n^2/(n^2 - 4)$$

when n is one of the numbers 3, 4, 5 or 6.

Further work discovered two important facts: the frequency of the light of a spectral line can always be expressed as the difference between two quantities, and is always related to a universal constant: a value that does not change. But why?

Lively atoms and quantum leaps

The Danish physicist Niels Bohr found the answer in 1913 by refining Ernest Rutherford's model of the atom (see page 281). Rutherford had a central nucleus surrounded by a spherical cloud of electrons. In Bohr's model, the electrons don't just whizz around the nucleus in any space available; they can only occupy fixed orbitals which are clearly defined. Each orbital is associated with a level of energy. You could think of them as being rather like the rungs of a ladder: an electron can only be on a rung, not in a space between rungs. These fixed energy levels were the key to understanding the coloured lines of emission spectra – and subsequently to unpicking the fabric of the material world by explaining how atoms interact.

A statue of Max Planck, who first stated that energy comes in 'quanta' or small packets.

Bohr explained that heating a substance provides energy to the atoms, making them agitated. An electron can then jump between energy levels. If it jumps down a level (that is, closer to the nucleus), it emits energy. This energy is of a fixed value for each particular movement between orbitals. The energy is thrown out as a photon, a tiny parcel of light. The photon is what produces the coloured light when gases are heated and it's how a neon light bulb produces light.

This is where the 'quantum' bit comes in. Max Planck had already demonstrated in 1900 that energy is emitted in packets of a fixed size – tiny quantities called quanta. Bohr recognized that this applied to his jumping electrons. The electron jumping between orbitals is performing a 'quantum leap'. (A quantum leap is commonly used to mean a huge change, but its original meaning is a tiny jump of a subatomic distance.)

From A to B via nowhere

While the leap is tiny in real terms, its impact on physics is massive. The electron, remember, can only exist in one of the allowed orbitals – it can't occupy any other position. This means it can't move between them by passing through the intervening space: it must jump instantaneously from one level to another. It must stop existing in one place and begin existing in another place at the same moment. This is only the beginning of the strangeness that quantum theory revealed.

Particles and waves

Louis de Broglie

We usually visualize an atom as a central lump of nucleus containing protons and neutrons, and then electrons in their allowed orbitals whizzing around it. Even if we know they are energy, we think of the electrons as little dots (it's how they are drawn in all the textbooks). But electrons turned out not to behave as we would expect a particle to behave.

The French physicist Louis de Broglie proposed in 1924 that electrons behave like both a wave and a particle at the same time. This seems to confound all expectations about the behaviour of electrons, and possibly other particles. It can be demonstrated, and was later, by an experiment.

Imagine a fork and a plate of peas. There is enough space between the prongs of the fork for a pea to pass through. If you roll peas towards the fork, they will either go through one of the gaps or bounce off a prong. If they go through, they will end up lined up behind the gaps. The English scientist Thomas Young performed a similar experiment in 1801, shining light through two slits on to a screen behind. But instead of just bands of light directly behind the gaps, he found interference patterns, which he claimed demonstrated that light is a form of wave.

A much more precise version of the experiment can be carried out using electrons. The apparatus consists of a plate with two slits in it arranged in front of a screen. A beam of electrons is fired towards the plate. An electron hitting the screen produces a spot of light. You would expect electrons to go through one slit or the other and pile up on the other side in line with one of the slits (or miss the slits entirely). But that's not what happens.

The electrons do not end up in two places on the screen, but produce a series of bands, some of which are behind parts of the plate with no slit – so the electrons, fired in a straight line perpendicular to the plate, should not be able to get there. The only explanation for the pattern was astonishing: the electrons were in some way behaving like waves and the pattern on the screen is an interference pattern, caused by peaks and troughs of the waves either reinforcing or cancelling each other out. Physicists struggled to explain it.

Here and there

The Austrian physicist Erwin Schrödinger suggested that the moving electrons smear out, so that they no longer behave like a single point particle. He produced equations to model the behaviour of these indefinite electrons.

But German physicist Max Born disagreed. His suggestion was bizarre, but it has not been refuted by any experimental evidence. He explained the wave pattern as a probability wave. The size of the wave at any location predicts the likelihood that the electron will be found there. Where the wave is large (a peak) there is a high probability that the electron will be found at that location. Where the wave is small (a trough) there is a low probability that the electron will be found there. The result is that we can't ask 'Where is the electron now?' because this can't be known with any certainty – all we can measure is the probability wave. So the question we can ask is: 'If I were to look here, what is the probability that I would find the electron there?' Schrödinger's equations can be used to predict the distribution of electrons, so if we fire lots of electrons we can use

Max Born

the equations to predict their distribution at different locations – but we can't ever say where any particular electron will be (or is).

Pinning down possibilities

It might sound as though we can never know the location of anything, but that's not quite true. Bohr pointed out that when we measure or observe something, its position is fixed. In the double-slit experiment, each electron's position is fixed at the point where it hits the screen as we can see traces that are produced by actual electrons. The difficulty is that we can't predict anything other than the probability of a location. And – perhaps worst of all – observing something changes it.

When we observe or measure a particle it is forced to relinquish all its other possible locations and be in the one place (or state) at which it is measured or observed. If detectors are used to determine which slit the electrons went through in the double-slit experiment, the interference patterns disappear. If only slightly reliable detectors are used, the interference pattern is reduced. Even worse, if the beam is slowed down so that just a single electron is fired and detected at once, the interference pattern still builds up over a sequence of electrons. The uncanny implication is that the electron 'knows' it is being watched and acts accordingly, or it 'knows' whether there are two slits open, even though it only goes through one. This has

WANTED, DEAD AND ALIVE: SCHRÖDINGER'S CAT

Schrödinger proposed a famous thought-experiment in 1935 to highlight what he considered the absurdity of some aspects of quantum theory – that a particle could exist in multiple states until observed. He suggested that a cat could be constrained in a box with a vial of poisonous gas, a radioactive source and a Geiger counter. If the Geiger counter detected that the radioactive source had decayed, a hammer would break the vial of poison and the cat would die. According to quantum theory, the state of the cat (living or dead) is not fixed until an observer looks in the box. So the cat is both alive and dead until then.

Quantum uncertainty works at the level of particles, rather than the level of cats, but in this case the fate of the cat depends on the state of a particle (whether or not radioactive decay has occurred), with its implications scaled up by the mechanical set-up of the experiment. Schrödinger wanted to show how silly this state of affairs is: the cat is clearly dead or alive whether or not it has been observed. But is it? Or is the cat its own observer? Perhaps this only works on inanimate objects.

not been fully explained, though there are theories that attempt to explain it.

Even the pinning down implied in measuring something doesn't lead to certainty about its state. Werner Heisenberg explained in his Uncertainty Principle, or Indeterminacy Principle, that we can't discover both the precise momentum of a particle and its precise position. The more accurately we measure one, the less we can know of the other. If you tried to measure the position of an electron by bouncing a photon off it and measuring the angle and speed of reflection of the photon, the observation would be faulty. The impact with the photon might impart momentum to the electron, or the electron might have moved on from the position it was in when the photon bounced off it and so the measurement of position will already be out of date by the time it is registered. Any interaction with a particle to measure it will change its state in some way.

Einstein's opposition

Not everyone was convinced by the arguments of quantum mechanics. One of the more famous dissenters was Albert Einstein (see page 216). Declaring that God does not play dice with the universe, he rejected the idea of uncertainty. He argued not that quantum theory was necessarily wrong, but that it was incomplete.

Entangled particles

One of the strangest predictions from the equations of quantum mechanics is entanglement. Two particles become entangled if they are very close together and their properties become linked. Changing a property of one particle automatically changes the corresponding property of the other. For example, if one had clockwise spin and the other had counter-clockwise spin, changing the spin of the first particle would change that of the second – it would have to, so that they remained of opposite spin. Of course, the direction of spin is not fixed until it's measured; then the possibilities collapse and it has to have one direction or the other. Measuring one particle then necessarily fixes the state of the entangled particle.

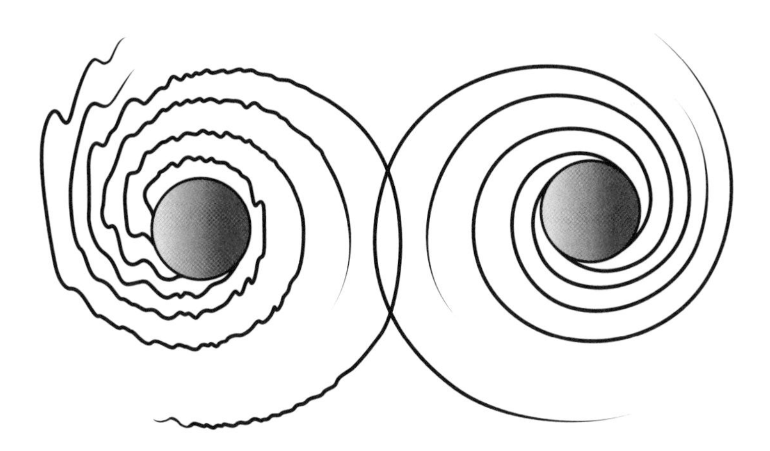

The bizarre thing is that the particles stay entangled even if separated. We could send one particle to the outer edge of the universe and the properties of the other should still change correspondingly – and instantaneously.

Einstein found this so ridiculous he referred to it as 'spooky action at a distance'. He agreed that the maths shows entanglement should exist, but that it couldn't be 'spooky'. He suggested another model

QUANTUM TELEPORTATION

It's too early to teleport objects, but in 2012 scientists successfully 'teleported' the state of a photon between two of the Canary Islands off the coast of Africa. They began by entangling two photons in La Palma, then sent one by laser to another island, Tenerife, nearly 150km (90 miles) away. Then they entangled a third photon with that in La Palma. Finally, when the quantum state of the third photon changed on La Palma, the state of the photon on Tenerife changed to reflect it immediately, with no delay at all.

This was not the first demonstration of quantum teleportation. The concept was proved in 2010 by a team in China over a distance of 16km (10 miles). The trial in the Canary Islands was significant in that it was far enough to show that it could be used to communicate with a satellite in Low Earth Orbit. In theory, quantum entanglement could be harnessed for instantaneous communication – but that's a long way off.

in which the state of each particle was determined at the start and only discovered when one was measured, not altered by its measurement. It seemed impossible to tell which theory was right: by measuring the properties of a particle, do we fix them or only discover them? Without an experimental dimension, there was stalemate between the two rival ideas.

In 1972, the American physicists John Clauser and Stuart Freedman built a machine to test quantum entanglement and found that it does actually exist. The results were confirmed by other experimental physicists elsewhere: quantum entanglement seems to be a genuine phenomenon.

Quantum future

Quantum theory seems to turn on its head a lot that we feel we know intuitively about the universe, but this doesn't mean it's wrong. The Copernican theory of a heliocentric solar system did the same. So far, experimental results have not found faults in quantum theory and a great deal of mathematics supports it. Its biggest remaining gap is that it is hard to reconcile with gravity. There are many ways in which people hope to put quantum mechanics to work, such as building super-fast tiny computers and possibly instantaneous means of communication. We are right at the beginning of the quantum age.

Picture credits

Alamy Stock Photo: 12, 16, 225 (National Geographic Creative), 274, 282, 297

Aleksejs Polakovs: 29

Aleksey Stemmer: 30

Getty Images: 20 (The LIFE Picture Collection), 39 (Bettmann Archive), 40 (The LIFE Picture Collection), 44 (Alinari), 48 (The LIFE Picture Collection), 66, 160, 166 (Corbis)

NASA: 21, 24, 129, 183, 184 (JPL/Space Science Institute), 191 (Bill Ingalls), 256 (The LIFE Picture Collection), 270 (De Agostini)

The Royal Society: 158

Science & Society Picture Library: 46 (Kodak Collection)

Science Photo Library: 59, 179, 232, 271, 290 (Detlev van Ravenswaay)

Shutterstock: 5, 6, 8, 11, 14, 15, 17, 18, 26, 28, 37, 41, 52, 53, 54, 55(b), 57(x2), 62, 63, 64, 65, 68, 69, 71, 73, 75(t), 78, 82, 83, 84, 85, 87, 98, 111, 114, 116, 119, 121, 122, 123(x2), 126, 127, 130, 142, 145, 146, 149, 151, 152, 156, 157, 159, 165, 168, 170, 172, 173, 175, 177, 180, 181, 190, 192, 195, 196, 199, 202, 204, 214, 216, 218, 221, 227, 234, 236, 243, 245, 246, 247, 248, 249, 250, 251, 259, 260, 262, 263, 265, 267, 269, 273, 280, 288, 289, 292, 293, 299

Wellcome Library, London: 88, 89, 91, 92, 93, 95, 97, 107, 108, 112, 125, 132, 134, 135, 141, 161, 162, 163, 167, 193, 198, 201, 206, 209, 210, 212, 222, 224, 239, 241, 277, 286, 295

www.apatrimonio.com: 27

WWW.SHOCK.CO.BA: 47

Diagrams by David Woodroffe: 25, 33, 51, 67, 72, 75(b). 80, 81, 101, 102, 105, 143, 148, 150, 154, 228, 230, 231, 238, 253, 279, 283, 285(x2), 301